化学工程与工艺专业英语

Professional English for Chemical Engineering and Technology

主审　王　军

主编　杨许召　张盈盈　韩敬莉

郑州大学出版社

图书在版编目(CIP)数据

化学工程与工艺专业英语／杨许召，张盈盈，韩敬莉主编. — 郑州：郑州大学出版社，2022. 8

ISBN 978-7-5645-8841-0

Ⅰ. ①化… Ⅱ. ①杨…②张…③韩… Ⅲ. ①化学工程 - 英语 - 高等学校 - 教材②化工过程 - 工艺学 - 英语 - 高等学校 - 教材 Ⅳ. ①TQ02②TQ06

中国版本图书馆 CIP 数据核字(2022)第 111266 号

化学工程与工艺专业英语

策划编辑	袁翠红	封面设计	苏永生
责任编辑	杨飞飞	版式设计	苏永生
责任校对	陈 思	责任监制	凌 青 李瑞卿

出版发行	郑州大学出版社	地 址	郑州市大学路40号(450052)
出 版 人	孙保营	网 址	http://www.zzup.cn
经 销	全国新华书店	发行电话	0371-66966070
印 刷	广东虎彩云印刷有限公司		
开 本	787 mm×1 092 mm 1 / 16		
印 张	15	字 数	465 千字
版 次	2022 年 8 月第 1 版	印 次	2022 年 8 月第 1 次印刷

| 书 号 | ISBN 978-7-5645-8841-0 | 定 价 | 39.00 元 |

本书如有印装质量问题,请与本社联系调换。

作者简介

杨许召,男,河南省汝州市人,中共党员,郑州轻工业大学材料与化学工程学院教学副院长,副教授,博士,河南省高等学校青年骨干教师。主要从事化学工艺、精细化工(表面活性剂、离子液体、化妆品等)的研究,相关研究工作发表于 *Fluid Phase Equilibria*、*Journal of Chemical Thermodynamics*、化工学报等,被评为"2016 年度中国洗涤用品行业中青年先进科技工作者"。近年来,主持/参与国家级、省部级项目 5 项;发表 SCI、EI、中文核心论文 70 余篇,授权国家发明专利 10 余件,获河南省科技进步三等奖 3 项,出版著作 7 部。

张盈盈,女,河南省安阳市人,中共党员,郑州轻工业大学材料与化学工程学院化学工程与工艺系副主任,讲师,博士,研究方向为新型离子液体用于 CO_2 分离过程的热力学分析及其在天然提取物、化妆品研制中的应用,相关研究工作发表于 *Applied Energy*、*Renewable and Sustainable Energy Reviews*、化学进展、中国科学化学、化工学报,主持/参与国家级、省部级项目 5 项,发表 SCI、EI 论文 10 余篇。

韩敬莉,女,河南省许昌市人,中共党员,郑州轻工业大学材料与化学工程学院化学工程与工艺系教师,讲师,博士,研究方向为离子液体预测型热力学模型及其在分离过程强化中的应用,相关研究工作发表于 *AIChE Journal*、*ACS Sustainable Chemistry & Engineering*、*Green Energy & Environment*、*Fluid Phase Equilibria*,授权国家发明专利 4 件,主持/参与国家级、省部级项目 4 项,发表 SCI、EI 论文 10 余篇。

前　言

随着经济全球化浪潮的涌起和我国精细化工产业的不断发展，社会和企业对化学工程与工艺人才素质的需求也在不断发展和变化，那些既能掌握化学工程与工艺专业技能和知识，又能熟练掌握化学工程与工艺专业英语的技术人才将受到社会和企业的青睐。为适应新时期社会经济变化对化学工程与工艺专业人才素质的需求，根据作者多年的教学经验，在原有校内讲义的基础上，收集大量最新专业资料，经过修改、补充，编写此书。

本书力图将科技英语的基本特点、构词法、翻译技巧和写作技巧介绍给读者，目的是让读者对科技英语有一个基本了解。本书在编写过程中努力体现以下特点：①本书所选内容尽量贴近专业实际，通俗易懂；②本书所选内容适应性强，覆盖面宽，可提高学生科技英语阅读水平和专业技能水平；③本书对课文中出现的生词、技术术语等都进行了音标注释和释义，便于学生阅读及理解。

本书分为六个部分，第一部分主要讲述科技英语的构词、翻译和写作，第二部分主要讲述四大基础化学、单元操作、传递过程和化工热力学，第三部分主要讲述化工过程模拟、分析和设计，第四部分主要讲述化工安全、环保和工程伦理，第五部分主要讲述表面化学和表面活性剂，第六部分主要讲述精细化学品，如清洗剂、颜料染料、化妆品和香精香料。本书第二部分第二单元和第五、六部分由郑州轻工业大学材料与化学工程学院杨许召老师编写，第一、四部分和第二部分第一单元由郑州轻工业大学材料与化学工程学院张盈盈老师编写，第三部分由郑州轻工业大学材料与化学工程学院韩敬莉老师编写。全书由郑州轻工业大学材料与化学工程学院王军教授主审。

本书得到了河南省高等教育教学改革研究与实践重点项目（2021SJGLX198）的资助。同时，本书的出版得到了郑州轻工业大学教务处的鼎力支持，在此深表谢意。

本书可作为化学工程与工艺及相关专业的专业英语教材，也可作为从事化工技术研究、开发、生产、管理的科研人员和工程技术人员的参考书。

本书涉及内容广泛，因编者水平有限，书中疏漏和不妥之处在所难免，恳请广大读者提出宝贵意见，以便完善。

编　者

2022 年 2 月

Content

Part 1

Introduction of English for Science and Technology

Unit 1   Introduction

After completing this unit, you should be able to:

1. Describe the definition and application of English for Science and Technology.
2. List the four catagories of science and technology words.
3. List the lexical characteristics, syntactic characteristics and rhetoric characteristics of EST.

Lesson 1   English for Science and Technology

1. Introduction to English for Science and Technology (EST)

English for Science and Technology (EST) generally refers to English used in scientific publications, papers, textbooks, technical reports and academic lectures, etc. It is used to describe the physical and natural phenomena, their processes, properties, characteristics, laws and application in productive activities.

As an outcome of the rapid development of science and technology after World War Two, EST initially emerged in the 1950s. Since the 1970s, together with the shift development of science and technology as well as the popularity of the English language, "EST has developed into an important variety of modern English in many countries", as pointed out by Qian Sanqing. Due to its main functions of statement, description, exposition, definition, classification, instruction, comparison, exemplification, inference and reasoning, EST has achieved its own language characteristics that contribute to the formal, concise, precise, impersonal and economical style of scientific documents.

2. The Lexical Characteristics of EST

The high-level profession and preciseness of ST materials are tactfully achieved by such lexical characteristics of EST as the frequent use of science and technology (ST) words, the replacement of verb phrases by verbs and the extensive use of abstract nouns and descriptive adjectives.

2.1 *The Frequent Use of ST Words*

Althoughfunction words and general words constitute the largest part of EST vocabulary, the frequent use of ST words in EST and the complicated way that they vary in formation, meaning and use still form a remarkable feature of EST vocabulary.

Generally, ST wordsfall into 4 categories according to the way that they are different in formation, meaning and use:

(1) Pure ST words such as *hydroxide, diode, promethazine and isotope* etc. These words mostly composed of Latin or Greek morphemes are monosemic and professionally used in a special field.

(2) Semi ST words such as *frequency, density, energy, magnetism*, etc. Compared to Pure ST words, these words are also monosemic but more commonly and frequently used in fields of different professions.

(3) Common ST words such as *feed, service, ceiling, power, operation, work*, etc. These are specialized common words carrying different meanings in fields of different professions. For example, the word feed with the basic meaning of to give food to a person or an animal can be used in different fields with different meanings such as *to supply water, to provide electricity, to deliver, to load, cutting feed, power source*, ect. Such polysemous common ST words are freely collocated with other words and are most widely and frequently used in fields of various professions.

(4) Built ST words such as *microbicid, waterleaf, medicare*, which are built through different ways of word building such as *affixation, compounding, blending, acronyms*, etc.

Built words are much more frequently used in EST than in general English to achieve the conciseness and preciseness of scientific documents. Such extensive use of various types of ST words seems to be a challenge for EST users; however, an awareness of the monosemy of pure and semi ST words, the specialization of common words and the polysemy of specialized common words together with knowledge of word-building will eventually lead to a good command of ST words.

2.2 *The Extensive Use of Abstract Nouns and Descriptive Adjectives*

In EST, there exist a large number of abstract nouns, which are used to indicate means, existence, tools as well as the results and states of actions, behaviors and movements. This kind of nouns mostly have the same roots with general verbs or adjective from which they are derived. For example,

insulate-insulation, expand-expansion, stable-stability, humid-humidity, etc.

Similarly, in EST also appear quite a quantity of descriptive adjectives used to describe the state, feature, degree, size and shape of naturalphenomona and matters. These adjective are mostly derived from verbs and nouns using suffixes such as

-ac/iac, -al, -ar, -ato, -eal, -ed, -ic- ible/able, -ing, -ing, -ive, -oid, - ose, ous, -y.

In addition, abstract nouns and descriptive adjectives are collocated to indicate highly professional existences. From such examples as

favorable prognosis, bacterial infection, systematic disorder, etc.

It can be seen the lexical collocation of abstract nouns and descriptive adjectives is another striking feature of EST vocabulary. The reason behind the fact that ST articles prefer such abstract nouns and the descriptive adjective is that the process, result, state, specialty and feature of existences in nature can be precisely and impersonally explained by abstract nouns and descriptive adjectives.

2.3 *The Replacement of Verb Phrases by Verbs*

A large number of English verbs and their corresponding verb phrases share the same meaning in a certain language context. For example,

absorb-take in, examine-look into, rotate-turn around, utilize-make good use of, etc.

Comparatively, single verbs of limited use are more monosemous and more formal, while verb phrases are more colloquial, synomemous and more widely used in spoken English. For example, *absorb* shares the same meaning with *take in* both meaning assimilate, but the latter flexibly carries different meanings of *understand, deceive, shorten* when used in different informal contexts. In order to avoid the ambiguity or colloquialism as well as to achieve the formality and preciseness of ST literature, formal verbs are preferentially used instead of verb phrases in EST.

3. The Syntactic Characteristics of EST

The accuracy, conciseness and objectivity of EST documents are also achieved by some syntactic characteristicsof EST, such as the extensive use of postpositive attributive, non-predicative verbs, passive voice as well as long and complicated sentences.

3.1 *Common Use of Postpositive Attributive*

Postpositive attributives commonly appearing in ordinary English are more frequently used in EST because of the preciseness requirement of ST documents. Generally, the use of postpositive attributives in EST can be realized through the following five structures:

(1) Prepositional phrase

e. g. *The explanations of the earth's magnetic field are generally accepted.*

(2) Adjectives or adjective phrases

e. g. *All radiant energy has wavelike characteristics, analogous to those of waves.*

(3) Adverbs

e. g. *The force upward equals the force downward so that the balloon stays at the level.*

(4) Participles

e. g. *In 1983, there were only 200 computers connected to the Internet.*

(5) Attributive Clauses

e. g. *The loads a structure is subjected are divided into dead loads, <u>which include the weights of all parts of the structure</u>, and live loads, <u>which are due to the weights of people, moveable equipment</u>.*

Besides, encouragingly used in EST are some postpositive attributives of more complicated structure, which are actually combinations of the the above postpositive attributives of simple structure. For example, in the sentence

Hand is the body part <u>at the end of a person's arm</u> that includes the fingers and thumb, <u>used for picking up, holding, and touching things</u>.

The scientific definition of *hand* is precisely and compactly presented by a combination of prepositional phrase, attributive clause, and participle phrase.

3.2 *Extensive Use of the Passive Voice*

Another prominent characteristic of EST syntactic structure is the extensive use of passive voice. At least one third of predicative verbs in EST documents are used in the passive voice. The extensive use of passive sentences meets the requirements of ST materials for objectivity, compactness and coherence. Firstly, the recounting and reasoning of ST works demand the objectivity of expression, and the use of the passive voice instead of the active voice helps to create a sense of objectivity. Secondly, in ST materials, the object of action is of greater importance than the subject, the actor who performs the action, and the use of the passive voice helps the object hold the head or the prominent position of the sentence. Additionally, in many cases, the prominence of the object of action will contribute to the compactness and coherence of ST literature. The following example will illustrate the advantages of passive voice in EST.

The Harry Diamond Laboratories performed early advanced development of the Arming Safety Device (ASD) for the Navy's 5-in guided projectile. The early advanced development was performed in two phrases. In phrase 1, the ASD was designed, and three prototypes were fabricated and tested in the laboratory. In phrase 2, the design was refined, 35 ASD's and a large number of explosive mockups were fabricated, and a series of qualification tests was performed. The qualification tests ranged from laboratory tests to drop tests and gun firing. The design was further refined during and following the qualification tests. The feasibility of the design was demonstrated.

The use of 8 passive clauses in the passage not only highlights the main contents being explained, but also creates objectivity, compactness and coherence of the whole passage.

3.3 *Wide Use of Non-predicative Verbs*

The non-predicative verbs usually taking three forms, namely "to + V" (the Infinitive), "V-ing" (the Present Participle or the Gerund) and "V-ed" (the Past Participle) are not limited by a subject or inflected by categories such as *tense, aspect, mood, number, gender, and person*, while at the same time play different roles of nouns, adjectives, or adverbs. Such flexible use of non-finite verbs creates the objectivity, conciseness, flexibility, and preciseness of expression, and thereby helps to avoid the subjectivity in meaning or the abundance in structure in ST materials. Therefore, non-predicative verbs are far more frequently used in EST than in general English to serve the purpose of explaining ideas impersonally, concisely and precisely. Consider the following sentence:

They use new and high technologies and non-public ownership mechanisms to enliven, transform and upgrade traditional industries, forcefully promoting the attainment of the objectives of "upgrading old enterprises, nurturing new industries and building large enterprises".

The sentence includes an infinitive phrase *to enliven, transform and upgrade traditional industries* and a participle phrase *forcefully promoting the attainment of the objectives of upgrading old enterprises, nurturing new industries and building large enterprises*, of which the latter is more complicated in that it further employs the gerund phrase *upgrading old enterprises, nurturing new industries and building large enterprises* as the object of the complicated prepositional phrase that modify *the attainment*, the object of the participle *promoting*.

3.4 *More Long and Complicated Sentences*

Science and Technology is the study of the development, distribution, structure and function of the living things in the outside world, which lie together in interrelated, paradoxical movements. To present the complicated relationships among the existences, EST documents greatly depends on the logic thinking that resort to the linguistic form-long and complicated sentences consisting of clauses and phrases that are mutually conditioned. Consider the following long and complicated sentence that is employed to describe the scientific possibility.

With the advent of the space shuttle, it will be possible to put an orbiting solar power plant in stationary orbit 24 000 miles from the earth that would collect solar energy almost continuously and convert this energy either directly to electricity via photovoltaic cells or indirectly with flat plate or focused collectors that would boil a carrying medium to produce steam that would drive a turbine that then in turn would generate electricity.

This long sentence involves a prepositional phrase *With the advent of the space shuttle* as the adverbial and the infinite phrase *to put an orbiting solar power plant...* as the real subject, which covers four encircled attributive clauses to modify *the orbiting solar power plant*. The long sentence of complicated structure expresses the complicated situation of putting a special solar power plant in a clear, logic and precise way.

4. The Rhetoric Characteristics of EST

The objectivity, compactness and coherence of ES materials can be further enhanced by the rhetorical features of EST, such as the limited use of tenses, the simple rhetorical choices, the common use of abnormal sentences, and the deliberate use of subjunctive mood and imperatives.

4.1 *The Limited Use of Tenses*

ES writings aims to objectively state the facts, describe the process, and illustrate the features and functions, most of which are of university, frequency and particularity. Therefore, in EST, more commonly used are general statements in simple tenses, mostly the simple present tense and simple past tense, to create timeless notions. Take the passage as an example.

In bear making. Yeast cells break down starch and sugar (present in cereal grains) to form alchol; the froth, or head, of the bear results from the carbon dioxide gas that the cells produce. In simple terms, the living cells rearrange chemical elements to form new productions that they need to live and reproduce. By happy coincidence in the process of doing so they help make a popular beverage.

Five verbs used in present tense, *break down, result from, produce, rearrange, help*, appear in the short passage of 3 sentences.

4.2 *The Less Use of Rhetorical Devices and Complex Formation Style*

Due to its function of stating facts and laws, EST seldom employ such rhetorical devices as metaphor, personification, hyperbole, etc commonly used in literature English. On the contrary, to directly explain the objective world, ST writers tend to adopt simple formational style of writing. For example, they often use the prefixion-statement pattern like the sentence

Sea water can be used for a supply of potable (or drinkable) water if it can be separated from the salt dissolved in it.

The main information is put in the head of the sentence, so that the reader can intuitively grasp the main point and understand the author's ideas immediately.

4.3 *The Frequent Use of Omission, Inversion and Separation*

In ST articles frequently appear incomplete, reversed, or fragmented sentences. The existence of such "abnormal" sentences is the rhetoric requirement more than the grammatical requirement. Specially, omission often helps to save space, making the presentation more concise and compact. e. g.

All bodies consist of molecules and molecules of atoms (consist is omitted after the second molecules).

Inversion often helps to make a common sense more eye-catching and prominent, the co-text more closely linked to and the description more vivid. e. g.

The most important of the materials in our bodies are theproteins (the subject the protein is inversed with the predicative are the most important).

Separation helps to enables the overall structure of the sentence more symmetric and balanced. e. g.

Thus, it would be correct to say that the distance to the sun, from where we are on the earth, is about 1 million walking days (the distance is separated from is about 1 million walking days).

4.4 *The Deliberated Use of Subjunctive Mood*

The description of common sense, the proposition of ideas, the discussion of problems and the deduction of formulas often involve a variety of prerequisites and conditions. In order to avoid ambiguity, the use of the subjunctive is always encouraged in EST. Additionally, many authors are willing to adopt subjunctive mood to express their humility and leave room for caution. For example,

If a pound of sand were broken up and turned into atomic energy, there would by enough power to supply the whole of Europe for a few years.

The use of subjective mood in the sentence avoids ambiguity in meaning and also leads to a polite and smooth tone for expressing the result of a supposed condition.

4.5 *The Tactful Use of Imperatives*

Imperatives are often used in EST materials, especially used in operation specifications, work procedures, and precaution matters to provide instructions, suggestions, advice and commands. Let us read the passage that presents the instructions for the process of an experiment.

Fill a test-tube half full of water and heat it nearly to boiling point. Support the tube on a stand and allow it to cool. Take the temperature every minute. Stir carefully with a glass rod. Record the readings you obtain, and plot them on a graph of temperature against time. Repeat this with a tube half-full of crystals. Allow the solid to melt. Heat the liquid to 100 ℃, fix the tube on the stand and allow it to cool. Record the results as before and plot them.

The use of the imperatives in the passage not only helps to make the article refined but also expresses the author's friendly suggestions to the reader. Additionally, each imperative beginning with a different verb adds to the variation of the writing, thus avoiding dull monotony. Imagine what kind of article it will become if you is put in the head of each sentence.

EST displays its own unique features in its evolution and development process. It is undoubtedly helpful for ST workers to understand the characteristics of EST that distinguish it from ordinary or literature English. However, it is also worth noting that the characteristics of EST are relative rather than absolute. In other words, the characteristics of EST also exist in ordinary or literature English but become more prominent in EST, and on the other hand, EST often borrows literary words and sentences or rhetorical devices to enhance the effect of language.

Words and Expressions

lexical [ˈleksɪkl]　*adj.* 词汇的

postpositive [poʊstˈpɒzɪtɪv]　*adj.* 置于词后的, 词尾的

Hydroxide [haɪˈdrɑːksaɪd]　*n.* 氢氧化物

attributive [əˈtrɪbjətɪv]　*n.* 定语

diode [ˈdaɪoʊd]　*n.* 二极管

prepositional [ˌprepəˈzɪʃənl]　*adj.* 前置词的, 介词的

promethazine [proʊˈmeθəziːn]　*n.* 异丙嗪

clauses [klɔːz]　*n.* 条款; 子句, 条

isotope [ˈaɪsətoʊp]　*n.* 同位素

gerund [ˈdʒerənd]　*n.* 动名词

Latin [ˈlætn] n. 拉丁语; adj. 拉丁语的

rhetoric [ˈretərɪk] n. 修辞, 修辞技巧; 花言巧语

Greek [griːk] n. 希腊语; adj. 希腊语的

deliberate [dɪˈlɪb(ə)rət] adj. 故意的, 蓄意的, 存心的

morphemes [ˈmɔːfiːmz] n. 语素, 词素

subjunctive [səbˈdʒʌŋktɪv] adj. 虚拟的

monosemic [mɒnəʊˈsiːmɪk] n. 单义的

imperatives [ɪmˈperətɪv] n. 祈使语气, 命令

polysemous [ˌpɑːliˈsiːməs] adj. (一词) 多义的

beverage [ˈbev(ə)rɪdʒ] n. (除水以外的) 饮料

microbicid [ˈmaɪkrəʊbɪk] n. 微生物制剂

metaphor [ˈmetəˌfɔr] n. 隐喻, 暗喻

conciseness [kənˈsaɪsnəs] n. 简明, 简洁; 切除

personification [pərˌsɑnəfɪˈkeɪʃ(ə)n] n. 拟人; 人格化

preciseness [priˈsaisnis] n. 准确, 一丝不苟

hyperbole [haɪˈpɜrbəli] n. 夸张, 夸张法

prognosis [prɑːgˈnoʊsɪs] n. 预断, 预后, 预测

omission [oʊˈmɪʃ(ə)n] n. 省略, 遗漏, 疏忽, 删除

synomemous [sɪˈnɒnɪməs] n. 同义词

Inversion [ɪnˈvɜrʃ(ə)n] n. 倒置, 颠倒, 倒转

ambiguity [ˌæmbɪˈgjuːəti] n. 歧义; 一语多义

Subjunctive [səbˈdʒʌŋktɪv] adj. 虚拟的, 虚拟语气的

colloquialism [kəˈləʊkwiəˌlɪz(ə)m] n. 口语; 俗语

tactful [ˈtæktfl] adj. 圆滑的, 得体的, 不得罪人的

syntactic [sɪnˈtæktɪk] adj. 句法的

Notes

Such extensive use of various types of ST words seems to be a challenge for EST users; however, an awareness of the monosemy of pure and semi ST words, the specialization of common words and the polysemy of specialized common words together with knowledge of word-building will eventually lead to a good command of ST words. 参考译文: 如此广泛地使用各种类型的科技词汇似乎对科技英语用户来说是一个挑战; 然而, 对纯科技词汇和半科技词汇的单义、普通词汇的专门化和专门化普通词汇的多义以及构词知识的认识最终将导致对科技词汇的良好掌握。

The reason behind the fact that ST articles prefer such abstract nouns and the descriptive adjective is that the process, result, state, specialty and feature of existences in nature can be precisely and impersonally explained by abstract nouns and descriptive adjectives. 参考译文: 科技论文之所以偏爱抽象名词和描述性形容词, 是因为抽象名词和描述性形容词可以准

确、客观地解释自然界存在的过程、结果、状态、特点和特征。

Exercises

1. Put the following into Chinese

Navigating instrument	Machine-shaping
Skin-deep	Quick-change
Sound-absorbing material	Standard-cubic-feet
State-of-the-art	Non-uniform

2. Change the following into passive voice.

We made a hole in a cork and pushed into it a narrow glass tube. Then we pushed this into the neck of a bottle which we had filled with colored water. When we did this, some of the colored water went up into the tube. We marked the level of the colored water in the tube.

Unit 2　Word Formation Processes

Before reading the text below, try to answer following questions:

1. Can you give the way of creating English new words?
2. Can you give the examples of new words of EST?

Lesson 1　Word Formation

There are some processes in creating English new words, such as a) affixation, b) folk etymology, c) compounding, d) abbreviation, e) acronyms, f) borrowing, g) blending, h) clipping, i) back-formation. Besides, there are also found the double word formation processes, such as, j) folk etymology + compounding, k) compounding + affixation, m) blending + affixation, n) clipping + blending. The result showed that the most productive process of creating English new words was affixation.

1. Affixation

Affixation has some types, such as prefix, suffix, infixes, and circumfixes. Then, in the findings, the types of affixation found are prefix, suffix, and circumfixes.

1.1 *Prefix*

For example, *autocyclic*. In the word "*autocylic*", the stem iscylic. It undergoes to affixation because it is added by prefix *auto-*. Some other examples are *biomethane, cycloheptane, cyclohexane*, etc.

1.2 *Suffix*

For example, *accretor*. In the word "*accretor*", the stem is accrete. It undergoes to affixation because it is added by suffix -*or*. Some other examples are *Aftonian, arcticized, bustler*, etc.

1.3 *Circumfixes*

For example, *afrofuturism*. In the word "*afrofuturism*", the stem is future. It undergoes to cirumfixes because it is added by prefix *afro-* and suffix -*ism*. Some other examples are *afrofuturist, anti-unionism*.

2. Borrowing

Borrowing is how to borrow the words from other language without any changing. The borrowing process which found in the findings arekinara, emoji, Naqada, mabuhay, angpow. The word "kinara" is borrowed from Swahili which means "candleholder". Besides, the word "Naqada" is borrowed from Arabic which means "the site of archaeological in Egyptian governorate of Qena". The word "mabuhay" is borrowed from Tagalog which means "greeting". Then, the word "angpow" is borrowed from Chinese. From the explanation above, it can be concluded that those new words are borrowed from another language without any changing.

3. Folk Etymology

Folk etymology is a little bit same with borrowing process. Folk etymology appeared because of historical story. It is because the speaker has different interpretation of the form. Therefore, the speaker changes the form or the pronounciation. However, in borrowing process, it is just borrowed witout any changing. The examples of folk etymology process which found are *apastron, falcial, ironice, fleishig*, etc. The word "apastron" is made from (apo-+ancient Greek στρον star), falcial (falx + ial), ironice (ironicus + classical Latin-ē), fleishig (from Yiddish fleyshik). From the explanation above, it can be seen that those new words are coming by borrowing from another language or by adding another language, such as classical Latin or ancient Greek for instance.

4. Compounding

Besides, the new words which are made by compounding process are *battleground, audio dub, aussieland, batchmate, hackboat*, etc. The word "battleground" is made from (battle + ground), audio dub (audio+dub), aussieland (Aussie+land), batchmate (batch+mate), hackboat (hack+boat). From the explanation above, it can be seen that those new words are coming by combining two words become a word.

5. Abbreviation

Furthermore, the new words which are made by abbreviation process are *CABG, Lw, Rt. Hon, SCBU, MRS*, etc. CABG is "Coronary Artery Bypass Graft", Lw is "Long wave", Rt. Hon is "Right Hon", SCBU is "Special Care Baby Unit", MRS is "Magnetic Resonance Spectroscopy". From the explanation above, it can be seen that it should be spoken letter by letter which is called as abbreviation.

6. Acronyms

Besides, the new words which are madeby acronyms process are *Captcha, Osha, ISA, YOLO*, etc. Captcha is coming from Completely Automated Public Turing test to tell Computers and Humans Apart, Osha is coming from Occupational Safety and Health Administration, ISA is coming from Industry Standard Architecture, YOLO is coming from You Only Live Once. Those are called as acronyms because the words can be read without speaking it letter by letter.

7. Blending

The blending processes which found in the findings are*cybercast, digipak, irone, chugger, backronym, brunello*, etc. The word "cybercast" is coming from cyber+broadcast. First, the word "broadcast" is clipped become "cast". Then, it combines with cyber. It can be concluded that blending has two processes which are clipping and compounding. Then it blends into a word. Some other examples are *digipak, irone, clicktivist*, etc.

8. Clipping

The new words which are created from clipping process are *dom, disco, demo, syst, Scandi*, etc. The word "dom" is dominus, the word "disco" is discotheque, the word "syst" is system, the word "Scandi" is Scandinavian. Those words are created by cutting the back which is called as final clipping. Therefore, it can be concluded that final clipping is the dominant type of clipping in creating a new word.

9. Back-formation

Back-formation is a little bit same with clipping. However, in back-formation, it can change the part of speech. While, in clipping, it does not change the part of speech. The new word which are created by back-formation process is gank. It is coming from "gangster". From the word "gank", it can be seen that the back is deleted and it changes into another one.

10. Double Word-formation Process

Double word-formation process is how to combine two derivational processes into a word. In the findings, the double word-formation processes found are:

10.1 *Folk Etymology + Compounding*

The new words can be created by folk etymology + compounding process. It means that there are two processes which are folk etymology and compounding. For example, historiosophy. It is coming from historio (from Greek) +sophy (from Greek). It can be seen that those word are coming from Greek and it combines into a word.

10.2 *Compounding + Affixation*

The new words can be created by compounding + affixation. It means that the new words are created by two processes. For example, *live-blogging, oversighting, beatboxer, dayworker, hardrocker*, etc. The word "live-blogging" is coming from live+blog+ing. The word "oversighting" is coming from over+sight+ing. The word "beatboxer" is coming from beat+box+er. The word "dayworker" is coming from day+work+er. The word "hardrocker" is coming from hard+rock+er.

10.3 *Blending+Affixation*

Furthermore, the new words can also be created by blending+affixation. For example, syntagmatics. It is coming from syntax+pragmatic+s. The word "syntax+pragmatic" is created by blending. Then, it is added by "s" as the suffix affixation.

10.4 *Clipping + Blending*

Then, clipping+blending process can also create a new word. For example, the word "d-line". It is coming from defensive line. It can be seen that the word defensive stands for "d". It can be seen that it is clipping process, then it blends with word "line".

Words and Expressions

affixation　[ˌæfɪkˈseɪˌʃən]　*n.* 附加法,加词缀法

afrofuturism　[ˈæfrouˈfjuːtʃərˌzəm]　*n.* 非洲未来主义

etymology　[ˌetɪˈmɑːlədʒi]　*n.* 词源,词源学,民间词源

afrofuturist　[ˈæfrouˈfjuːtʃərˌst]　*n.* 非洲未来主义者

compounding　[kəmˈpaundɪŋ]　*n.* 混合物,组合物

unionism　[ˈjuːniənˌzəm]　*n.* 工会主义,联邦主义

abbreviation　[əˌbriːviˈeɪˌʃn]　*n.* 缩写,缩写词

emoji　[ˌɪˈmoudʒi]　*n.* 表情符号

acronyms　[ˈækrənˌmz]　*n.* 首字母缩略词

angpow n. a gift in the form of money

borrowing　[ˈbɔːrouˌŋ]　*n.* 借用的言语(或思想)

Swahili　[swəˈhiːli]　*n.* 斯瓦希里语

blending　[ˈblendˌŋ]　*n.* 融合,混合,组合

Tagalog　[təˈgɑːlɒg]　*n.* 他加禄语(通行于菲律宾群岛)

clipping　[ˈklɪpˌŋ]　*n.* 剪报,剪下物

archaeological　[ˌɑrkiəˈlɑdʒˌk(ə)l]　*adj.* 考古学的

back-formation　[ˈbæk fɔːrmeˌʃn]　*n.* 逆构词

apastron　[əpɑːstˈrən]　*n.* 远星点

prefix　[ˈpriːfˌks]　*n.* 前缀,前置代号

falcial　[ˈfælʃl]　*adj.* 镰的,镰状的;网络镍的

suffix　[ˈsʌfˌks]　*n.* 后缀

ironice　[aˌɪrɑːnˌk]　*adj.* 说反话的,讽刺的

infix　[ˈɪnfˌks]　*n.* 中缀,中加成分

fleishig　[ˈflaˌʃˌgz]　*adj.* 认作是肉

circumfixes　[ˈsɜrkəmˌfleks]　*n.* 音调符号,扬抑符

gangster　[ˈgæŋstər]　*n.* 歹徒,匪徒,土匪

autocylic　[ˈɔtouˈsaˌlˌk]　*adj.* 自动循环的

bustler　[ˈbʌslər]　*n.* 喧闹者,活跃者

Notes

1. Aftonian:阿夫顿阶,阿夫顿间冰期,北美更新世的一个间冰期,是内布拉斯加冰期与堪萨冰期之间的一个气候温暖期。

2. Naqada:奈加代,上埃及地名,位于尼罗河西岸。

3. Qena:基纳,位于尼罗河东岸,基纳干谷(Wadi Qena)流入红海的交汇点,是来往上埃及和红海之间的主要交通航道,穿梭乐蜀和红河的旅客必经此地,因此经济繁荣。

Exercises

1. Put the following into English

保水的   耗时的   热处理   人造卫星   上层建筑

光敏性的   碳氢化合物   磺基甜菜碱   P-N 结   PM2.5

2. Please translate the following sentence into Chinese, and list the new words and its word formation process.

The development of research on synthesis of Butyl Butyrate and the catalysis of sulfonic acid, inorganic salt, solid superacid, heteropoly acid, enzyme catalysis compound are summarized.

Unit 3　Writing

After completing this unit, you should be able to:
1. Describe the definition of scientific writing and classical format of scientific paper.
2. If reader want to know the content of the paper, which part should be reading first?

Lesson 1　Scientific Writing

1. Introduction to Scientific Writing

Scientific writing is a technical form of writing that is designed to communicate scientific information to other scientists. Depending on the specific scientific genre—a journal article, a scientific poster, or a research proposal, for example—some aspects of the writing may change, such as its purpose, audience, or organization. Many aspects of scientific writing, however, vary little across these writing genres.

Preparing publishable manuscripts is an important. Scientific writing skills can be developed through a long process of training and experience. The first step in developing a manuscript is to focus on a subject or problem that might be of significant interest to colleagues in the field. Next, the prospective writer must do a detailed survey of the relevant literature, the results of which will help him or her decide whether to actually write about the topic. Since the primary goal is to get the manuscript published, the writer should bear a specific journal in mind and write in accordance with the guidelines of that publication. He or she must also consider general ethics and scientific rules during the writing process.

Learning how to assess and use scientific sources, how to relate the collected information to the manuscript, and how to write in good scientific form are all important.

2. Title Page

This page contains the title of the article, which should be short, descriptive, informative and catchy. The authors' names, highest academic degrees, and institutional affiliation are also noted, as are the names of the department(s) and institution(s) where the work was done. The addresses of the corresponding author and of the author responsible for reprints should also appear on this page.

2.1 *Authorship*

Authorshipand the number of the authors is an important issue. Only those individuals who have actively participated in planning and conducting the work should be listed as authors. Every author shares equal responsibility for the scientific content of the manuscript. Authors are also responsible for protecting patient privacy, and for obtaining permission from the original author if diagrams, photographs, illustrations, or long quotations from other sources are used. They are also responsible for arranging copyright transfer. Honorary or gift authorship, although practiced by some, is not generally approved. The following criteria are specified for authorship:

Ⅰ. Conception and design, or analysis and interpretation of data;

Ⅱ. Drafting the article or revising it critically for important intellectual content;

Ⅲ. Final approval of the version to be published.

2.2 Abstract and Key Words

The Abstract is an important section of the manuscript. For many readers, the decision whether or not to read the full text rests on the quality of this summary. Maximum information about the manuscript must be provided in a minimum number of words, since many journals have a set word count for abstracts. The tremendous number of articles and difficulty selecting the pertinent literature from this pile led to the concept of the "structured abstract". In this type of abstract, the information is divided under subheadings. Such reporting is expected to supply the reader with uniform and standardized information. The argument against this format is that it reduces creativity. Structured abstracts are requested by some journals in order to facilitate more rapid identification of articles, to pave the way for more precise computerized literature searches, and to help editors during the peer review process. The Key Words should appear in alphabetical order, and be in accordance with the subject headings list in Index.

The credibility of a manuscript and the size of the audience it reaches are, to some extent, reflected by the number of times it is cited in other articles. The Title, Abstract, and Key Words are all important elements for catching the reader's attention.

3. Introduction

The Introduction is the first part of the classical IMRAD format. This section familiarizes the reader with the problem, question, or hypothesis that the paper will discuss. The known aspects of the subject are summarized as orientation to the topic. The last portion of this section should tell what is unknown or problematic about the specific subject. The purpose of the study is also conveyed in the Introduction.

4. Material and Methods

This section describes the entire process that was used in effort to answer the problem described in the Introduction. To start, the author should outline the material and equipment used, and their sources. The study design (descriptive study, orthogonal experiment, prospective or retrospective study) should also be stated. Materials and Methods may be divided into subsections to impart a better understanding of the process that was undertaken. No results are reported in this section.

5. Results

In this section, the author reports all the concrete evidence. The structure of Results should matchthat of Material and Methods, in that there are corresponding results for each method that has been described. The writer must also report the characteristics, the duration of the trial, and any differences between what was planned and what was actually done, including reasons for the differences.

6. Discussion

The Discussion is the final part of the main text. Answers and solutions to the questions and problems posed in the Introduction are given. This section should include not only the positive and desired or expected results, but should also note any deficiencies, discrepancies, and limitations of the study, including possible sources of bias. The results are compared with findings in the pertinent literature, and similarities and differences are highlighted. The Discussion must include the extent to which findings and conclusions can be generalized to other field, with implications for applicability and exclusions.

7. References

All references should be listed either in the order they appear in the text or alphabetically, based on the last name of the first author. The list must be complete and accurate. If the same author has written more than one of the references, then the list order should be based on year of publication, from earliest to most recent. Personal communications and unpublished data may also be referenced. Reporting of personal communication must be kept to a minimum, and should be used only to convey data not found in other sources. The authors must find and read all the references, and cross-referencing should be avoided. Each item in the reference list must be cited in the text.

Words and Expressions

catchy　[ˈkætʃi]　*adj.* 悦耳易记的

prospective　[prəˈspektɪv]　*adj.* 预期的, 潜在的

quotations　[kwoʊˈteɪʃənz]　*n.* 引语, 引文, 引用, 引述

retrospective　[ˌretrəˈspektɪv]　*adj.* 回顾的, 涉及以往的

honorary　[ˈɑːnəreri]　*adj.* 名誉的, 荣誉的

orthogonal　[ɔːˈθɒɡənəl]　*adj.* 正交的

gift authorship　酬庸作者

discrepancies　[dɪˈskrɛpənsiz]　*n.* 差异, 不一致

hypothesis　[haɪˈpɑːθəsɪs]　*n.* 假设, 假说, 猜想

scenarios　[sɪˈnɛrioʊz]　*n.* 设想, 预测, 梗概

descriptive　[dɪˈskrɪptɪv]　*adj.* 描述性的

genre　[ˈʒɑːnrə]　*n.* 体裁, 类型

Notes

IMRAD 论文各部分,观察性研究和实验性研究论文的正文通常(不是必须)分为几个部分,以前言(引言)、方法、结果和讨论作为各部分的标题。这种结构被称为"IM-RAD"。IMRAD 结构直接反映了科学发现的过程。

Exercises

1. Download three papers from different journals in your own major and list the structure of the paper. Compare the structure of these paper.

2. Make a judgement on the following description right or wrong:

(1) The title is the single most important phrase of a scientific document.

(2) The title tells readers what the document is. If your title is inexact or unclear, many people for whom you wrote the document will also read it.

(3) The purpose of a abstract is to summarizes the main sections of the paper, States the purpose, findings, and impact of the work.

(4) Besides presenting an analysis of the key results in the conclusion sections, you also give a future perspective on the work.

(5) Methods = What data were accumulated. Results = How the data were accumulated.

(6) Results = Data presentation ("Experiments showed that…"). Discussion = Data interpretation ("Experiments suggest that…")

Unit 4   Translation

After completing this unit, you should be able to answer the following question:

1. What is translation and why we need it?

2. What should be taken in account in translation?

3. What are the four level of translation?

4. What are the advantages and disadvantages of the two translation methods, which are human translation and machine translation?

Lesson 1   Human Translation

1. Introduction to Translation

Translation is the transmittal of written text from one language into another. Although the terms translation and interpretation are often used interchangeably, translation refers to the written language, and interpretation to the spoken word. Translation is the action of interpretation of the meaning of a text, and subsequent production of an equivalent text, also called a translation, that communicates the same message in another language. The text to be translated is called the source text, and the language it is to be translated into is called the target language.

The need for translation has existed since time immemorial and translating important literary works from one language into others has contributed significantly to the development of world culture. Ideas and forms of one culture have constantly moved and got assimilated into other cultures through the works of translators.

The history of translation is related to the history of the often invisible cross cultural interactions of the world. Ideas and concepts from the East notably India, China and Iraq have influenced the Western culture since as early as sixth century B. C. when trade ties were first established between India and the Mediterranean countries.

Translation must take into account constraints that include context, the rules of grammar of the two languages, their writing conventions, and their idioms. A common misconception is that there exists a simple word-for-word correspondence between any two languages, and that translation is a straightforward mechanical process. A word-for-word translation does not take into account context, grammar, conventions, and idioms.

2. The Level of Translation

There are four levels more or less consciously for translation: ①the source language text

019

level, the level of language, where we begin and which we continually (but not continuously) go back to; ②the referential level, the level of objects and events, real or imaginary, which we progressively have to visualize and build up, and which is an essential part, first of the comprehension, then of the reproduction process; ③the cohesive level, which is more general, and grammatical, which traces the train of thought, the feeling tone (positive or negative) and the various presuppositions of the source language text. This level encompasses both comprehension and reproduction: it presents an overall picture, to which we may have to adjust the language level; ④the level of naturalness, of common language appropriate to the writer or the speaker in a certain situation. Again, this is a generalized level, which constitutes a band within which the translator works, unless he is translating an authoritative text, in which case he sees the level of naturalness as a point of reference to determine the deviation-if any-between the author's level he is pursuing and the natural level. This level of naturalness is concerned only with reproduction. Finally, there is the revision procedure, which may be concentrated or staggered according to the situation. This procedure constitutes at least half of the complete process.

2.1 *The Textual Level*

This is the level of the literal translation of the source language into the target language, the level of the translationese you have to eliminate, but it also acts as a corrective of paraphrase and the parer-down of synonyms. So a part of your mind may be on the text level whilst another is elsewhere. Translation is pre-eminently the occupation in which you have to be thinking of several things at the same time.

2.2 *The Referential Level*

Whether a text is technical or literary or institutional, you have to make up your mind summarily and continuously: what it is about, what it is in aid of, what the writer's peculiar slant on it is. For each sentence, when it is not clear, when there is an ambiguity, when the writing is abstract or figurative, you have to ask yourself: What is actually happening here? and why? For what reason, on what grounds, for what purpose? Can you see it in your mind? Can you visualize it? If you cannot, you have to "supplement" the linguistic level, the text level with the referential level, the factual level with the necessary additional information (no more) from this level of reality, the facts of the matter.

The referential goes hand in hand with the textual level. All languages have polysemous words and structures which can be finally solved only on the referential level, beginning with a few multi-purpose, overloaded prepositions and conjunctions, through dangling participles to general words.

2.3 *The Cohesive Level*

Beyond the second factual level of translating, there is a third, generalized, level linking the first and the second level, which you have to bear in mind. This is the "cohesive" level; it follows both the structure and the moods of the text: the structure through the connective words

(conjunctions, enumerations, reiterations, definite article, general words, referential synonyms, punctuation marks) linking the sentences, usually proceeding from known information (theme) to new information (theme; proposition, opposition, continuation, reiteration, conclusion—for instance—or thesis, antithesis, synthesis) . Thus the structure follows the train of thought; ensures that a colon has a sequel, that ulterieur has a later reference; that there is a sequence of time, space and logic in the text.

2.4 *The Level of Naturalness*

The level of naturalness of natural usage is grammatical as well as lexical (i. e. , the most frequent syntactic structures, idioms and words that are likely to be appropriately found in that kind of stylistic context) , and, through appropriate sentence connectives, may extend to the entire text,

In all "communicative translation", whether you are translating an informative text, a notice or an advert, "naturalness" is essential. That is why you cannot translate properly if the traget language is not your language of habitual usage. That is why you so often have to detach yourself mentally from the source language text; why, if there is time, you should come back to your version after an interval.

3. The General Law of Translation

The general law of translation is as follows,

Ⅰ. That the Translation should give a complete transcript of the ideas of the original work.

Ⅱ. That the style and manner of writing should be of the same character with that of the original.

Ⅲ. That the Translation should have all the ease of original composition.

4. Approaches to Translation

There are two approaches to translating (and many compromises between them) : ①you start translating sentence by sentence, for say the first paragraph or chapter, to get the feel and the feeling tone of the text, and then you deliberately sit back, review the position, and read the rest of the source language text; ②you read the whole text two or three times, and find the intention, register, tone, mark the difficult words and passages and start translating only when you have taken your bearings. Which of the two methods you choose may depend on your temperament, or on whether you trust your intuition (for the first method) or your powers of analysis (for the second) . Alternatively, you may think the first method more suitable for a literary and the second for a technical or an institutional text. The danger of the first method is that it may leave you with too much revision to do on the early part, and is therefore time-wasting. The second method (usually preferable) can be mechanical; a transitional text analysis is useful as a point of reference, but it should not inhibit the free play of your intuition.

It is the duty of a translator to attend only to the sense and spirit of his original, to make himself perfectly master of his author's ideas, and to communicate them in those expressions which he judges to be best suited to convey them. It has, on the other hand, been maintained, that, in order to constitute a perfect translation, it is not only requisite that the ideas and sentiments of the original author should be conveyed, but likewise his style and manner of writing, which, it is supposed, cannot be done without a strict attention to the arrangement of his sentences, and even to their order and construction. According to the former idea of translation, it is allowable to improve and to embellish; according to the latter, it is necessary to preserve even blemishes and defects; and to these must likewise be superadded the harshness that must attend every copy in which the artist scrupulously studies to imitate the minutest lines or traces of his original.

Words and Expressions

immemorial [ˌɪməˈmɔːriəl] *adj.* 古老的, 远古的

figurative [ˈfɪɡjərətɪv] *adj.* 比喻的, 形象的

context [ˈkɑːntekst] *n.* 上下文, 语境

conjunctions [kənˈdʒʌŋkʃənz] *n.* 连词, 连接词

grammar [ˈɡræmər] *n.* 语法, 文法

enumeration [ˌnuːməˈreɪʃn] *n.* 列举事实, 逐条陈述

conventions [kənˈvenʃn] *n.* 习俗, 常规, 惯例

reiteration [riˌɪtəˈreɪʃn] *n.* 重覆, 反覆, 重说

idioms [ˈɪdiəmz] *n.* 习语, 成语, 惯用语

definite article [ˌdefɪnət ˈɑːrtɪkl] 定冠词

textual [ˈtekstʃuəl] *adj.* 文本的, 篇章的

general words [ˈdʒenrəl wɜːrdz] 例行文句, 全面性词句

cohesive [koʊˈhiːsɪv] *adj.* 有结合力的

punctuation marks [ˌpʌŋktʃuˈeɪʃn mɑːrks] 标点符号

presuppositions [ˌpriːsəpəˈzɪʃnz] *n.* 假定, 预设

Mediterranean [ˌmedɪtəˈreɪniən] *adj.* 地中海的

whilst [waɪlst] *conj.* 同时, 同 while(连词)

referential synonyms 指称同义词

Notes

Ideas and concepts from the East notably India, China and Iraq have influenced the Western culture since as early as sixth century B. C. when trade ties were first established between India and the Mediterranean countries.

参考译文: 早在公元前 6 世纪, 当印度和地中海国家之间首次建立贸易关系时, 来自东方的思想和观念, 尤其是印度、中国和伊拉克, 就已经影响了西方文化。

Exercises

1. Translate the following sentences into Chinese.

(1) Even though fossil fuel burning and various other activities release much larger quantities of CO_2, than anthropogenic release of CH_4, the impact of CH_4 vis a vis global warming is quite high because each molecule of CH_4 has 25 times more global warming potential (GWP) than a molecule of CO_2.

(2) One of the most perceptive observations on the human experience have come from George Santayana, the Spanish philosopher-essayist-poet-novelist who has famously said: Those who cannot remember the past are condemned to repeat it. The same thought has also been equally famously expressed in these words: history tends to repeat itself.

2. Download a scientific English paper and translate the abstract and introduction into Chinese.

Lesson 2   Machine Translation

1. Introduction to Machine Translation

The dream of a universal translation device goes back many decades, long before Douglas Adams's fictional Babel fish provided this service in The Hitchhiker's Guide to the Galaxy. Since the advent of computers, research has focused on the design of digital Machine Translation tools-computer programs capable of automatically translating a text from a Source Language to a Target Language. This has become one of the most fundamental tasks of artificial intelligence.

Machine Translation (MT) is an automated translation of text performed by a computer. It provides text translations based on computer algorithms without human involvement. With Machine Translation, source text is easily and quickly translated into one or more target languages. Usually fast and simple to use, MT engines represent a quick and easy, although not always the best solution for translation.

Machine Translation is not to be confused with Computer Assisted Translation (CAT). While CAT includes the use of different Machine Translation engines, their role is usually supportive to the human translator. On the other hand, Machine Translation is the sole product of the computer, although human reviewers might be included for quality assurance.

2. History of Machine Translation

The translation of natural languages by machine, first dream of in the seventeenth century, has become a reality in the late twentieth century. Computer programs are producing translations. The translations are not perfect, for that is an ideal to which no human translator can aspire, nor are they translations of literary texts, for the subtleties and nuances of poetry are beyond computational analysis, but translations of technical manuals, scientific documents, commercial prospectuses, administrative memoranda, medical reports. Machine Translation is not primarily an area of abstract intellectual inquiry but the application of computer and language sciences to the development of systems answering practical needs.

After an outline of basic features, the history of Machine Translation is traced from the pioneers and early systems of the 1950s and 1960s, the impact of the Automatic Language Processing Advisory Committee (ALPAC) report in the mid-1960s, the revival in the late 1970s, the appearance of commercial and operational systems in the 1980s, research during the 1980s, new developments in research in the 1990s, and the growing use of systems in the 1990s.

3. Things to Consider When Opting for Machine Translation

3.1 *Accelerated Translation Workflow*

Machine Translations are done much faster.

3.2 *Type of Content That is Being Translated*

Not all content is suitable for MT. However, content which is frequently changed or updated would demand high costs of human translation and MT might be the perfect solution.

3.3 *Not All Languages Provide the Same Quality of Results*

Machine Translation still provides the best results when translating from and to English, with several languages close behind: Spanish, French, German, Portuguese.

3.4 *Human Factor*

No matter the quality of MT, human professional translators are still the most reliable option. Are you translating generic phrases or specialized or niche texts? Before choosing MT, consider including a human reviewer in your translation process.

3.5 *Software*

Software, such as Bing and Google Translate (GT), is very useful for translating technical documents, particularly manuals.

The accuracy of MT strictly depends on 1) the language you are translating from-the greater the number of speakers of your language, the better the translation is likely to be 2) how much you modify the text of the original language before submitting it for translation. The second factor is crucial. Before submitting your text to GT, it is needed to make it more English for example by: ① changing the word order to reflect English word order; ② reducing the length of sentences; ③ replacing pronouns (e. g. it, one, them) with their respective nouns; ④ removing redundancy.

4. Typical Areas Where Google Translate May Make Mistakes in English

4.1 *Word Order*

GT's main difficulty is with word order, i. e. the position of nouns, verbs, adjectives, and adverbs. If in your language you put the verb before its subject, or if you put an indirect object before the direct object, then GT will not be able to create the correct English order (i. e. the reverse of the order in your language).

4.2 *Plural Acronyms*

In English, we say one CD but two CDs. Most other languages do not have a plural form for acronyms, and thus say two CD. GT is able to recognize this for common acronyms such as CD, DVD and PC, but not for very technical acronyms.

4.3 *Tenses*

GT sometimes changes the tense from the original. For example, you may use the future tense and GT will translate it into the present tense. In some cases, GT may be correct. For example, if in your language you say "I will tell him when I will see him", GT will correctly translate this as "I will tell him when I see him". This is because in a time clause, when takes the present and not the future. However, very occasionally GT makes mistakes when it changes tenses, so it is wise to check very carefully.

4.4 *Uncountable Nouns*

An uncountable noun is a noun that cannot be made plural and which cannot be preceded by a/an or one. For example information is uncountable. This means you cannot say an information, one information, two informations, several informations.

The problem with uncountable nouns is that the surrounding words (i. e. articles, pronouns and verbs) must also be singular. This means that if, for example, information is countable in your language, GT will probably make errors with the surrounding words, as highlighted in this example (note: the example is NOT in correct English):

These information are vital in order to understand xyz. In fact, they are so crucial that ...

4.5 *Very Specialized Vocabulary*

GT's dictionaries are huge but do not cover absolutely every word. If GT doesn't know a word, it will normally leave it in the original language.

4.6 *Words with More Than One Meaning*

GT generally manages to guess the right meaning when translating into English because it looks at the surrounding words (i. e. how words are collocated together). In any case, you need to check carefully that GT has translated with the meaning you intended.

4.7 *Strings of Words Used in Computer Terminology*

If you use English phrases such as status no-provider in your own original language, sometimes GT will modify these when "translating" and produce, for example, provider-no status. Essentially, you just need to check that any English in your source text has not been "translated" by GT.

4.8 *Names of People*

At the time of writing, GT tends to translate people's first names and sometimes surnames. This should not be a problem in manuals as names of people do not usually appear. In any case, be aware that GT makes some rather unexpected translations.

4.9 *Accents and Single Quotes*

Does your native language use accents? If it does, then read on. If you are, for example, French, then GT is helped considerably if you use the correct accents. Note how GT translates these two titles of a French medical paper in two ways depending on whether the accents are inserted or not. Interestingly, both translations would be possible, but one of the two might not reflect the author's real intention.

Words and Expressions

memoranda [ˌmɛməˈrændə]　*n.* 协议备忘录

Plural [ˈplurəl]　*adj.* 复数的,多元的

intellectual [ˌntəˈlektʃuəl]　*adj.* 智力的,脑力的

CD [ˌsiːˈdiː]　*n.* compact disc 光盘,光碟

inquiry [ˈɪnkwəri]　*n.* 调查,查询

DVD [ˌdiːviːˈdiː]　*n.* digital videodisc 数字光碟

Computer Assisted Translation　计算机辅助翻译

PC[ˌpiːˈsiː]　*n.* personal computer 个人计算机

Notes

The Hitchhiker's Guide to the Galaxy:银河系漫游指南,影片根据《银河系漫游指南》小说改编。主要人物是一个无家可归的地球人,一个到处为家的外星人,还有几位个性鲜明的角色。

Douglas Adams's fictional Babel fish:道格拉斯·亚当斯虚构的巴别塔鱼,是一条钻入高等智慧生物的耳朵后能够让宿主听懂全宇宙所有语言的"寄生鱼"。

Automatic Language Processing Advisory Committee:语言自动处理咨询委员会。1964年,为了对机译的研究进展作出评价,美国科学院成立了语言自动处理咨询委员会(ALPAC),开始了为期两年的综合调查分析和测试。ALPAC 于 1966 年 11 月公布了一个题为《语言与机器》的报告,该报告全面否定了机译的可行性,并建议停止对机译项目的资金支持。这份报告的公开发表给了正在蓬勃发展的机译当头一棒,各国的机译研究陷入了近乎停滞的僵局。

Exercises

1. Translate the following sentences into Chinese.

(1) The overall magnitude of the task being disconcertingly huge: capturing and safely se-

questering as much as 25 billion tonnes of CO_2 that is generated worldwide by anthropogenic activities every year, the environmental impacts of all decarbonizations processes will, inevitably, to be proportionately large and complex.

(2) Realising the limitations of sequestering fossil fuel CO_2, great hope has been pinned on renewable energy sources. The year 2008 has witnessed more funds to be allocated, globally, to renewable-based power generation than ever before.

2. Download a scientific English paper in your field and translate the Result and Discussion into Chinese.

Part 2 — Chemistry and Chemical Engineering

Unit 1   General Chemistry

After completing this unit, you should be able to:
1. Identify the differences between organic chemistry and inorganic chemistry.
2. Define the research category of physical chemistry.
3. Describe the nature of analytical chemistry.

Lesson 1   Inorganic Chemistry

1. Introduction to Inorganic Chemistry

The word organic refers to the compounds which contain the carbon atoms in it. So the branch of chemistry that deals with the study of compounds, which does not consist of carbon-hydrogen atoms in it, is called "Inorganic Chemistry. " In simple words, it is opposite to that of Organic Chemistry. The substances which do not have carbon-hydrogen bonding are the metals, salts, chemical substances, etc.

On this planet, there are known to exist about 100 000 number of inorganic compounds. Inorganic chemistry studies the behaviour of these compounds along with their properties, their physical and chemical characteristics too. The elements of the periodic table except for carbon and hydrogen, come in the lists of inorganic compounds.

2. Elements

The elements are broadly divided into metals, nonmetals, and metalloids according to their physical and chemical properties; the organization of elements into the form resembling the modem periodic table is accredited to Mendeleev.

A useful broad division of elements is into metals and nonmetals. Metallic elements (such as iron and copper) are typically lustrous, malleable, ductile, electrically conducting solids at about room temperature. Nonmetals are often gases (oxygen), liquids (bromine), or solids that do not conduct electricity appreciably (sulfur). The chemical implications of this classification should already be clear from introductory chemistry:

(1) Metallic elements combine with nonmetallic elements to give compounds that are typically hard, nonvolatile solids (for example sodium chloride).

(2) When combined with each other, the nonmetals often form volatile molecular compounds (for example phosphorus trichloride).

(3) When metals combine (or simply mix together) they produce alloys that have most of the physical characteristics of metals (for example brass from copper and zinc).

Some elements have properties that make it difficult to classify them as metals or non metals. These elements are called metalloids. Examples of metalloids are silicon, germanium, arsenic, and tellurium.

A more detailed classification of the elements is the one devised by Dmitri Mendeleev in 1869; this scheme is familiar to every chemist as the periodic table. Mendeleev arranged the known elements in order of increasing atomic weight (molar mass). This arrangement resulted in families of elements with similar chemical properties, which he arranged into the groups of the periodic table. For example, the fact that C, Si, Ge, and Sn all form hydrides of the general formula EH_4 suggests that they belong to the same group. That N, P, As and Sb all form hydrides with the general formula EH_3 suggests that they belong to a different group. Other compounds of these elements show family similarities, as in the formulas CF_4 and SiF_4 in the first group, and NF_3 and PF_3.

Mendeleev provided a spectacular demonstration of the usefulness of the periodic table by predicting the general chemical properties, such as the numbers of bonds they form, of unknown elements corresponding to gaps in his original periodic table. (He also predicted elements that we now know cannot exist and denied the presence of elements that we now know do exist, but that is overshadowed by his positive achievement and has been quietly forgotten.) The same process of inference from periodic trends is still used by inorganic chemists to rationalize trends in the physical and chemical properties of compounds and to suggest the synthesis of previously unknown compounds. For instance, by recognizing that carbon and silicon are in the same family, the existence of alkenes $R_2C=CR_2$ suggests that $R_2Si=SiR_2$, ought to exist too. Compounds with silicon-silicon double bonds (disila-ethenes) do indeed exist, but it was not until 1981 that chemists succeeded in isolating one.

The blocks of the periodic table reflect the identity of the orbitals that are occupied last in the building-up process. The period number is the principal quantum numberof the valence shell. The group number is related to the number of valence electrons.

The layout of the periodic table reflects the electronic structure of the atoms of elements. We can now see, for instance, that a block of the table indicates the type of subshell currently being occupied according to the building-up principle. Each period, or row, of the table corresponds to the completion of the s and p subshells of a given shell. The period number is the value of the principal quantum number n of the shell which according to the building-up principle is currently being occupied in the main groups of the table. For example, Period 2 corresponds to the n = 2 shell and the filling of the 2s and 2p subshells.

3. Bonding

The majority of inorganic compounds exist as solids and comprise ordered arrays of atoms, ions, or molecules. Some of the simplest solids are the metals, the structures of which can be described in terms of regular, space-filling arrangements of the metal atoms. These metal centres interact through metallic bonding, a type of bonding that can be described in two ways. One view is that bonding occurs in metals when each atom loses one or more electrons to a common "sea". The strength of the bonding results from the combined attractions between all these freely moving electrons and the resulting cations. An alternative view is that metals are effectively enormous molecules with a multitude of atomic orbitals that overlap to produce molecular orbitals extending throughout the sample.

Metallic bonding is characteristic of elements with low ionization energies, such as those on the left of the periodic table, through the d block, and into part of the p block close to the d block. Most of the elements are metals, but metallic bonding also occurs in many other solids, especially compounds of the d-metals such as their oxides and sulfides. Compounds such as the lustrous-red rhenium oxide ReO, and fool's gold (iron pyrites, FeS) , illustrate the occurrence of metalic bonding in compounds.

The familiar properties of a metal stem from the characteristics of its bonding and in particular the delocalization of electrons throughout the solid. Thus, metals are malleable (easily deformed by the application of pressure) and ductile (able to be drawn into a wire) because the electrons can adjust rapidly to relocation of the metal atom nuclei and there is no directionality in the bonding. They are lustrous because the electrons can respond almost freely to an incident wave of electromagnetic radiation and reflect it.

In ionic bonding ions of different elements are held together in rigid, symmetrical arrays as a result of the attraction between their opposite charges. Ionic bonding also depends on electron loss and gain, so it is found typically in compounds of metals with electronegative elements. However, there are plenty of exceptions: not all compounds of metals are ionic and some compounds of nonmetals (such as ammonium nitrate) contain features of ionic bonding as well as covalent interactions. There are also materials that exhibit features of both ionic and metallic bonding.

Ionic and metallic bonding are nondirectional, so structures where these types of bonding occur are most easily understood in terms of space-filling models that maximize, for example, the number and strength of the electrostatic interactions between the ions. The regular arrays of acorns, ions, or molecules in solids that produce these structures are best represented in terms of the repeating units that are produced as a result of the efficient methods of filling space.

4. Acid and Base

The original distinction between acids and bases was based, hazardously, on criteria of taste and feel: acids were sour and bases felt soapy. A deeper chemical understanding of their propertiesemerged from Arrheniuss (1884) conception of an acid as a compound that produced hydrogen ions in water. The definition due to Brønsted and Lowry focuses on proton transfer, and that due to Lewis is based on the interaction of electron pair acceptor and electron pair donor molecules and ions.

Acid-base reactions are common, although we do not always immediately recognize them as such, especially if they involve more subtle definitions of what it is to be an acid or base. For instance, production of acid rain begins with a very simple reaction between sulfur dioxide and water:

$$SO_2(g) + H_2O(l) \longrightarrow HOSO_2^-(aq) + H^+(aq)$$

This will turn our to be a type of acid-base reacrion. Saponification is the process used in soapmaking:

$$NaOH(aq) + RCOOR'(aq) \longrightarrow NaRCO_2(aq) + R'OH(aq)$$

This too is a type of acid-base reaction. There are many such reactions, and in due course we shall see why they should be regarded as reactions between acids and bases.

5. Oxidation and Reduction

Oxidation is the removal of electrons from a species; reduction is the addition of electrons. Almost all elements and their compounds can undergo oxidation and reduction reactions and the element is said to exhibit one or more different oxidation states. A large class of reactions of inorganic compounds can be regarded as occurring by the transfer of electrons from one species to another. Electron gain is called reduction and electron loss is called oxidation; the joint process is called a redox reaction. The species that supplies electrons is the reducing agent (or reductant) and the species that removes electrons is the oxidizing agent (or "oxidant"). Many redox reactions release a great deal of energy and they are exploited in combustion or battery technologies.

Because of the diversity of redox reactions it is often convenient to analyse them by applying a set of formal rules expressed in terms of oxidation numbers and not to think in terms of actual electron transfers. Oxidation then corresponds to an increase in the oxidation number of an element and reduction corresponds to a decrease in its oxidation number. If no element in a reaction undergoes a change in oxidation number, then the reaction is not redox. We shall adopt this approach when we judge it appropriate.

Words and Expressions

metalloids ['metlɔɪd] *n.* 准金属,类金属
electronegative [ˌɪlektrəʊ'negətɪv] *adj.* 带负电的
germanium [dʒɜːr'meɪniəm] *n.* 锗
saponification [səˌpɒnəfə'keɪʃən] *n.* 皂化反应
arsenic ['ɑːrsnɪk] *n.* 砷

reductant [rɪˈdʌktənt] *n.* 还原剂, 还原性介质

tellurium [teˈlʊriəm] *n.* 碲

redox [ˈredɔks] *n.* 氧化还原反应

brass [bræs] *n.* 黄铜

oxidant [ˈɑːksɪˌdənt] *n.* 氧化剂

lustrous [ˈlʌstrəs] *adj.* 有光泽的, 柔软光亮的

ammonium nitrate [əˈmoʊniəm ˈnaɪˌtreɪt] *n.* 硝酸铵

delocalization [diːˌloʊkələˌzeɪʃn] *n.* 离域作用

rhenium [ˈriːniəm] *n.* 铼

directionality [dɪˌrekʃəˈnælɪtɪ] *n.* 指向性, 定向性

phosphorus trichloride 三氯化磷

Notes

Dmitri Mendeleev:德米特里·门捷列夫,俄国科学家,发现并归纳元素周期律,依照原子量,制作出世界上第一张元素周期表,并据以预见了一些尚未发现的元素。

Arrheniuss:阿仑尼乌斯,瑞典化学家。提出了电解质在水溶液中电离的阿伦尼乌斯理论,研究了温度对化学反应速率的影响,得出阿伦尼乌斯方程。由于在物理化学方面的杰出贡献,被授予 1903 年诺贝尔化学奖。

Brønsted and Lowry:布朗斯特(Brønsted)和劳里(Lowry)提出的酸碱定义是:"凡是能够释放出质子（H^+）的物质,无论它是分子、原子或离子,都是酸;凡是能够接受质子的物质,无论它是分子、原子或离子,都是碱。

Exercises

1. Distinguish whether the following substances are inorganic compounds or not.

ammonium nitrate H_2SCH_2══CH_2 carbon tetrachloridecarbon dioxide

Ureacholine chloridesodium hydroxideglucosesulphuric acid

2. Matching the following word.

(1) battery technologies 电子对给体

(2) electron pair acceptor 金属键

(3) ionization energy 空间填充排布

(4) electron pair donor 亚电子层

(5) metalic bonding 电离能

(6) space-filling arrangements 双硅烷乙烯

(7) subshell 电子对受体

(8) disila-ethenes 电池技术

Lesson 2　Analytical Chemistry

1. The Introduction of Analytical Chemistry

Analytical chemistry is the branch of chemistry that deals with the analysis of different substances. Analytical chemistry involves the separation, identification, and the quantification of matter. It involves the use of classical methods along with modern methods involving the use of scientific instruments.

The output of analytical chemistry is high-quality measurements; the input is the conception of new tools, both physical and computational. The bridge from input to output is an infrastructure that allows computation, simulation, mechanical drawing, and fabrication-based on skills in electronics, optics, vacuum, and mechanical design-brought together to create new devices (sensors, instruments, and prototypes). The infrastructure and frame of mind needed for such activities are completely different from those required for, say, synthetic chemistry. Increasingly, analytical chemistry is a central part of branches of science other than chemistry. The subject is closer to some branches of engineering than to traditional chemistry, making the term analytical science an increasingly appropriate descriptor.

Analytical chemistry involves the following methods: The process of separation isolates the required chemical species which is to be analysed from a mixture. The identification of the analyte substance is achieved via the method of qualitative analysis. The concentration of the analyte in a given mixture can be determined with the method of quantitative analysis.

2. The Nature of Analytical Chemistry

Analytical chemistry is a measurement science consisting of a set of powerful ideas and methods that are useful in all fields of science, engineering, and medicine. Both qualitative and quantitative information are required in an analysis. Qualitative analysis establishes the chemical identity of the species in the sample. Quantitative analysis determines the relative amounts of these species, or analytes, in numerical terms. The data from the various spectrometers on the rovers contain both types of information. As is common with many analytical instruments, the gas chromatograph and mass spectrometer incorporate a separation step as a necessary part of the analytical process. With a few analytical tools, chemical separation of the various elements contained in the rocks is unnecessary since the methods provide highly selective information. In this text, we will explore quantitative methods of analysis, separation methods, and the principles behind their operation. A qualitative analysis is often an integral part of the separation step, and determining the identity of the analytes is an essential adjunct to quantitative analysis.

3. The Role of Analytical Chemistry

Analytical chemistry is applied throughout industry, medicine, and all the sciences. To illustrate, consider a few examples. The concentrations of oxygen and of carbon dioxide are determined in millions of blood samples every day and used to diagnose and treat illnesses. Quantities of hydrocarbons, nitrogen oxides, and carbon monoxide present in automobile exhaust gases are measured to determine the effectiveness of emission-control devices. Quantitative measurements of ionized calcium in blood serum help diagnose parathyroid disease in humans. Quantitative determination of nitrogen in foods establishes their protein content and thus their nutritional value. Analysis of steel during its production permits adjustment in the concentrations of such elements as carbon, nickel, and chromium to achieve a desired strength, hardness, corrosion resistance, and ductility. The mercaptan content of household gas supplies is monitored continually to ensure that the gas has a sufficiently obnoxious odor to warn of dangerous leaks. Farmers tailor fertilization and irrigation schedules to meet changing plant needs during the growing season, gauging these needs from quantitative analyses of plants and soil.

Quantitative analytical measurements also play a vital role in many research areas in chemistry, biochemistry, biology, geology, physics, and the other sciences. For example, quantitative measurements of potassium, calcium, and sodium ions in the body fluids of animals permit physiologists to study the role these ions play in nerve-signal conduction as well as muscle contraction and relaxation. Chemists unravel the mechanisms of chemical reactions through reaction rate studies. The rate of consumption of reactants or formation of products in a chemical reaction can be calculated from quantitative measurements made at precise time intervals. Materials scientists rely heavily on quantitative analyses of crystalline germanium and silicon in their studies of semiconductor devices whose impurities lie in the concentration range of 1×10^{-6} to 1×10^{-9} percent. Archaeologists identify the sources of volcanic glasses (obsidian) by measuring concentrations of minor elements in samples taken from various locations. This knowledge in turn makes it possible to trace prehistoric trade routes for tools and weapons fashioned from obsidian.

Many chemists, biochemists, and medicinal chemists devote much time in the laboratory gathering quantitative information about systems that are important and interesting to them. All branches of chemistry draw on the ideas and techniques of analytical chemistry. Analytical chemistry has a similar function with respect to the many other scientific fields listed in the diagram. Chemistry is often called the central science; its top center position and the central position of analytical chemistry in the figure emphasize this importance. The interdisciplinary nature of chemical analysis makes it a vital tool in medical, industrial, government, and academic laboratories throughout the world.

4. A Typical Quantitative Analysis

4.1 Choosing a Method

The choice is sometimes difficult and requires experience as well as intuition. One of the firstquestions that must be considered in the selection process is the level of accuracy required.

Unfortunately, high reliability nearly always requires a large investment of time. The selected method usually represents a compromise between the accuracy required and the time and money available for the analysis.

A second consideration related to economic factors is the number of samplesthat will be analyzed. If there are many samples, we can afford to spend a significant amount of time in preliminary operations such as assembling and calibrating instruments and equipment and preparing standard solutions. If we have only a single sample or a just a few samples, it may be more appropriate to select a procedure that avoids or minimizes such preliminary steps. Finally, the complexity of the sample and the number of components in the sample always influence the choice of method to some degree.

4.2 *Acquiring the Sample*

The second step in a quantitative analysis is to acquire the sample. To produce meaningful information, an analysis must be performed on a sample that has the same composition as the bulk of material from which it was taken. Whenthe bulk is large and heterogeneous, great effort is required to get a representative sample. Consider, for example, a railroad car containing 25 tons of silver ore. The buyer and seller of the ore must agree on a price, which will be based primarily on the silver content of the shipment. The ore itself is inherently heterogeneous, consisting of many lumps that vary in size as well as in silver content. The assay of this shipment will be performed on a sample that weighs about one gram. For the analysis to have significance, the composition of this small sample must be representative of the 25 tons (or approximately 22 700 000 g) of ore in the shipment. Isolation of one gram of material that accurately represents the average composition of the nearly 23 000 000 g of bulk sample is a difficult undertaking that requires a careful, systematic manipulation of the entire shipment.

Sampling is the process of collecting a small mass of a material whose composition accurately represents the bulk of the material being sampled. The collection of specimens from biological sources represents a second type of sampling problem. Sampling of human blood for the determination of blood gases illustrates the difficulty of acquiring a representative sample from a complex biological system. The concentration of oxygen and carbon dioxide in blood depends on a variety of physiological and environmental variables. For example, applying a tourniquet incorrectly or hand flexing by the patient may cause the blood oxygen concentration to fluctuate. Because physicians make life-and-death decisions based on results of blood gas analyses, strict procedures have been developed for sampling and transporting specimens to the clinical laboratory. These procedures ensure that the sample is representative of the patient at the time it is collected and that its integrity is preserved until the sample can be analyzed.

Many sampling problems are easier to solve than the two just described. Whether sampling is simple or complex, however, the analyst must be sure that the laboratory sample is representative of the whole before proceeding. Sampling is frequently the most difficult step in an analysis and the source of greatest error. The final analytical result will never be any more reliable than the reliability of the sampling step.

4.3 *Processing the Sample*

The third step in an analysis is to process the sample. Under certain circumstances, no sample processing is required prior to the measurement step. For example, once a water sample is withdrawn from a stream, a lake, or an ocean, the pH of the sample can be measured directly. Under most circumstances, we must process the sample in one of several different ways. The first step in processing the sample is often the preparation of a laboratory sample.

4.3.1 *Preparing a Laboratory Sample*

A solid laboratory sample is ground to decrease particle size, mixed to ensure homogeneity, and stored for various lengths of time before analysis begins. Absorption or desorption of water may occur during each step, depending on the humidity of the environment. Because any loss or gain of water changes the chemical composition of solids, it is a good idea to dry samples just before starting an analysis. Alternatively, the moisture content of the sample can be determined at the time of the analysis in a separate analytical procedure.

Liquid samples present a slightly different but related set of problems during the preparation step. If such samples are allowed to stand in open containers, the solvent may evaporate and change the concentration of the analyte. If the analyte is a gas dissolved in a liquid, as in our blood gas example, the sample container must be kept inside a second sealed container, perhaps during the entire analytical procedure, to prevent contamination by atmospheric gases. Extraordinary measures, including sample manipulation and measurement in an inert atmosphere, may be required to preserve the integrity of the sample.

4.3.2 *Defining Replicate Sample*

Most chemical analyses are performed on replicate samples whose masses or volumes have been determined by careful measurements with an analytical balance or with a precise volumetric device. Replication improves the quality of the results and provides a measure of their reliability. Quantitative measurements on replicates are usually averaged, and various statistical tests are performed on the results to establish their reliability.

4.3.3 *Preparing Solutions: Physical and Chemical Change*

Most analyses are performed on solutions of the sample made with a suitable solvent. Ideally, the solvent should dissolve the entire sample, including the analyte, rapidly and completely. The conditions of dissolution should be sufficiently mild that loss of the analyte cannot occur. Unfortunately, many materials that must be analyzed are insoluble in common solvents. Examples include silicate minerals, high-molecular-mass polymers, and specimens of animal tissue.

With such substances, we must follow the flow diagram to the box on the right and perform some rather harsh chemistry. Converting the analyte in such materials into a soluble form is often the most difficult and time-consuming task in the analytical process. The sample may require heating with aqueous solutions of strong acids, strong bases, oxidizing agents, reducing agents, or some combination of such reagents. It may be necessary to ignite the sample in air or oxygen or to perform a high-temperature fusion of the sample in the presence of various fluxes. Once the analyte is made soluble, we then ask whether the sample has a property that is proportional to analyte concentration and that we can measure. If it does not, other chemical steps may be necessary, to convert the analyte to a form that is suitable for the measurement step. For example, in the determination of manganese in steel, the element must be oxidized to MnO_4^- before the absorbance of the colored solution is measured. At this point in the analysis, it may be possible to proceed directly to the measurement step, but more often than not, we must eliminate interferences in the sample before making measurements, as illustrated in the flow diagram.

4.4 *Eliminating Interferences*

Once we have the sample in solution and converted the analyte to an appropriate form for measurement, the next step is to eliminate substances from the sample that may interfere with measurement. Few chemical or physical properties of importance in chemical analysis are unique to a single chemical species. Instead, the reactions used and the properties measured are characteristic of a group of elements of compounds. Species other than the analyte that affect the final measurement are called interferences, or interferents. A scheme must be devised to isolate the analytes from interferences before the final measurement is made. No hard and fast rules can be given for eliminating interference. This problem can certainly be the most demanding aspect of an analysis.

4.5 *Calibrating and Measuring Concentration*

All analytical results depend on a final measurement X of a physical or chemical property of the analyte. This property must vary in a known and reproducible way with the concentration c_A of the analyte. Ideally, the measurement of the property is directly proportional to the concentration, that is

$$c_A = kX$$

where k is a proportionality constant. With a few exceptions, analytical methods require the empirical determination of k with chemical standards for which c_A is known. The process of determining k is thus an important step in most analyses; this step is called a calibration.

4.6 *Calculating Results*

Computing analyte concentrations from experimental data is usually relatively easy, particularly with computers. These computations are based on the raw experimental data collected in the measurement step, the characteristics of the measurement instruments, and the stoichiometry of the analytical reaction.

4.7 *Evaluating Results by Estimating Reliability*

As the final step shows, analytical results are complete only when their reliability has been estimated. The experimenter must provide some measure of the uncertainties associated with computed results if the data are to have any value.

5. An Integral Role for Chemical Analysis: Feedback Control Systems

Analytical chemistry is usually not an end in itself but is part of a bigger picture in which the analytical results may be used to help control a patient's health, to control the amount of mercury in fish, to control the quality of a product, to determine the status of a synthesis, or to find out whether there is life on Mars. Chemical analysis is the measurement element in all of these examples and in many other cases. Consider the role of quantitative analysis in the determination and control of the concentration of glucose in blood.

Patients suffering from insulin-dependent diabetes mellitus develop hyperglycemia, which manifests itself in a blood glucose concentration above the normal concentration range of 65 mg/dL to 100 mg/dL. We begin our example by determining that the desired state is a blood glucose level below 100 mg/dL. Many patients must monitor their blood glucose levels by periodically submitting samples to a clinical laboratory for analysis or by measuring the levels themselves using a handheld electronic glucose monitor.

The first step in the monitoring process is to determine the actual state by collecting a blood sample from the patient and measuring the blood glucose level. The results are displayed, and then the actual state is compared to the desired state. If the measured blood glucose level is above 100 mg/dL, the patient's insulin level, which is a controllable quantity, is increased by injection or oral administration. After a delay to allow the insulin time to take effect, the glucose level is measured again to determine if the desired state has been achieved. If the level is below the threshold, the insulin level has been maintained, so no insulin is required. After a suitable delay time, the blood glucose level is measured again, and the cycle is repeated. In this way, the insulin level in the patient's blood, and thus the blood glucose level, is maintained at or below the critical threshold, which keeps the metabolism of the patient under control.

The process of continuous measurement and control is often referred to as a feedback system, and the cycle of measurement, comparison, and control is called a feedback loop. These ideas are widely applied in biological and biomedical systems, mechanical systems, and electronics. From the measurement and control of the concentration of manganese in steel to maintaining the proper level of chlorine in a swimming pool, chemical analysis plays a central role in a broad range of systems.

Words and Expressions

analyte [ˌænəˈlˌɪt] n. 分析物, 分析元素
interferences [ˌɪntərˈfɪrəns] n. 干扰物, 干涉
arathyroid [ˌpærəˈθaɪˌɔɪd] adj. 甲状旁腺附近的
diabetic mellitus 糖尿病
nickel [ˈnɪˌk(ə)l] n. 镍
hyperglycemia [ˌhaɪˌpərglaɪˈsimiə] n. 高血糖
chromium [ˈkroʊmiəm] n. 铬

blood glucose [blʌd 'gluːkoʊs] *n.* 血糖

mercaptan [mə'kæptæn] *n.* 硫醇

insulin-dependent diabete 胰岛素依赖型糖尿病

obnoxious [ɑb'nɑkʃəs] *adj.* 极讨厌的, 可憎的

gas chromatograph 气相色谱仪

unravel [ʌn'ræv(ə)l] *v.* 解开, 解体, 崩溃

mass spectrometer 质谱仪

crystalline ['krɪst(ə)lˌaɪn] *adj.* 结晶的, 晶状的

automobile exhaust gases 汽车尾气

volcanic [vɑl'kænˌk] *adj.* 火山的, 火山产生的

nerve-signal conduction 神经信号传导

interdisciplinary [ˌɪntər'dɪsˌplˌneri] *adj.* 多学科的, 跨学科的

handheld electronic glucose monitor 手持式电子血糖仪

fluxes [flʌks] *n.* 焊剂, 助熔剂

quantitative analytical measurements 定量分析测量

manganese ['mæŋgəniːz] *n.* 锰

stoichiometry [ˌstɔˌkiˈɒmətrˌ] *n.* 化学计量学

Exercises

Distinguish whether the following methods are analytical method or not.

nuclear magnetic resonance Liquid chromatography Titration FT-IR spectroscopy

Determination of saponification number   Water quality analysis

Lesson 3 Organic Chemistry

1. Introduction to Organic Chemistry

Organic chemistry is the study of the structure, properties, composition, reactions, and preparation of carbon-containing compounds. Most organic compounds contain carbon and hydrogen, but they may also include any number of other elements (e. g. , nitrogen, oxygen, halogens, phosphorus, silicon, sulfur) .

2. Alkylation of Enolates and Other Carbon Nucleophiles

Carbon-carbon bond formation is the basis for the construction of the molecular framework of organic molecules by synthesis. One of the fundamental processes for carbon-carbon bond formation is a reaction between a nucleophilic and an electrophilic carbon. Carbon nucleophiles, which are enolates, imine anions, and enamines can react with alkylating agents. Mechanistically, these are usually S_N2 reactions in which the carbon nucleophile displaces a halide or other leaving group with inversion of configuration at the alkylating group. Efficient carbon-carbon bond formation requires that the S_N2 alkylation be the dominantreaction. The crucial factors that must be considered include: ①the conditions for generation of the carbon nucleophile; ②the effect of the reaction conditions on the structure and reactivity of the nucleophile; and ③ the regio- and stereo-selectivity of the alkylation reaction. The reaction can be applied to various carbonyl compounds, including ketones, esters, and amides.

3. Reactions of Carbon Nucleophiles with Carbonyl Compounds

The reactions include some of the most useful methods for carbon-carbon bond formation: the aldol reaction, the Robinson annulation, the Claisen condensation and other carbon acylation methods, and the Wittig reaction and other olefination methods. All of these reactions begin with the addition of a stabilized carbon nucleophile to a carbonyl group. The product that is isolated depends on the nature of the stabilizing substituent (Z) on the carbon nucleophile, the substituents (A and B) at the carbonyl group, and the ways in which A, B, and Z interact to complete the reaction pathway from the addition intermediate to the product. Four fundamental processes are outlined below. Aldol addition and condensation lead to β-hydroxyalkyl or α-alkylidene derivatives of the carbon nucleophile. The acylation reactions follow the pathway that a group leaves from the carbonyl electrophile. In the Wittig and related olefination reactions, the oxygen in the adduct reacts with the group Z to give an elimination product. Finally, if the enolate has an α-substituent that is a leaving group, cyclization can occur. This is observed, for example, with enolates of α-haloesters.

4. Reduction of Carbon-Carbon Multiple Bonds, Carbonyl Groups, and Other Functional Groups

Reduction reactions are especially important in synthesis. Reduction can be accomplished by several broad methods including addition of hydrogen and/or electrons to a molecule or by removal of oxygen or other electronegative substituents. The most widely used reducing agents from a synthetic point of view are molecular hydrogen and hydride derivatives of boron and aluminum. The most widely used method for adding the elements of hydrogen to carboncarbon double bonds is catalytic hydrogenation. Except for very sterically hindered alkenes, this reaction usually proceeds rapidly and cleanly. The most common catalysts are various forms of transition metals, particularly platinum, palladium, rhodium, ruthenium, and nickel. Both the metals as finely dispersed solids or adsorbed on inert supports such as carbon or alumina (heterogeneous catalysts) and certain soluble complexes of these metals (homogeneous catalysts) exhibit catalytic activity. Depending upon conditions and catalyst, other functional groups are also subject to reduction under these conditions.

5. Concerted Cycloadditions, Unimolecular Rearrangements, and Thermal Eliminations

A certain reaction involves polar or polarizable reactants and proceed through polar intermediates and/or transition structures. One reactant can be identified as nucleophilic and the other as electrophilic. Carbanion alkylations, nucleophilic additions to carbonyl groups, and electrophilic additions to alkenes are examples of such reactions. The reactions to be examined in this chapter, on the other hand, occur via a reorganization of electrons through transition structures that may not be much more polar than the reactants. These reactions proceed through cyclic transition structures. The activation energy can be provided by thermal or photochemical excitation of the reactant(s) and often no other reagents are involved. Most of the transformations fall into the category of concerted pericyclic reactions, in which there are no intermediates and the transition structures are stabilized by favorable orbital interactions. These reactions can be classified into three broad types: cycloadditions, unimolecular rearrangements, and eliminations.

6. Organometallic Compounds of Group I and II Metals

The use of organometallic reagents in organic synthesis had its beginning around 1900 when Victor Grignard discovered that alkyl and aryl halides react with magnesium metal to give homogeneous solutions containing organomagnesium compounds. The "Grignard reagents" proved to be highly reactive carbon nucleophiles and are still very useful synthetic reagents.

Organolithium reagents came into synthetic use somewhat later, but are also very important for synthesis. The composition of the organolithium compounds is RLi or more accurately (RLi)$_n$. The organomagnesium compounds are usually formulated as RMgX, with X being a halide. The organometallic derivatives of Group I and II metals provide reactive carbon nucleophiles. Reactivity increases in the order Li < Na < K and MgX < CaX, but the lithium and magnesium reactions are by far the most commonly used. Organolithium and magnesium reagents react with polar multiple bonds, especially carbonyl groups, and provide synthetic routes to a variety of alcohols. Other electrophiles, such as acyl halides, nitriles, and CO_2 provide routes to ketones and carboxylic acids The Group II B organometallics derived from zinc, cadmium, and mercury are considerably less reactive. The carbon-metal bonds in these compounds have more covalent character than for lithium or magnesium reagents. Zinc, cadmium, and mercury are distinct from other transition metals in having a d10 shell in the +2 oxidation state and their reactions usually do not involve changes in oxidation state. Although organozinc and cadmium reagents react with acyl chloride, reactions with other carbonyl compounds require either Lewis acids or chelates as catalysts. These catalyzed reactions make organozinc reagents particularly useful in additions to aldehydes. The lanthanides and indium organometallics are usually in the +3 oxidation state, which are also filled valence shells, and have a number of specialized applications that depend on their strong oxyphilic character.

7. Carbon-Carbon Bond-forming Reactions of Compounds of Boron, Silicon, and Tin

Boron, silicon, and tin are at the metal-nonmetal boundary, with boron being the most and tin the least electronegative of the three. The neutral alkyl derivatives of boron have the formula

R_3B, whereas silicon and tin are tetravalent compounds, R_4Si and R_4Sn. These compounds are relatively volatile nonpolar substances that exist as discrete molecules and in which the carbon-metal bonds are largely covalent. By virtue of the electron deficiency at boron, the boranes are Lewis acids. Silanes do not have strong Lewis acid character but can form pentavalent adducts with hard bases such as alkoxides and especially fluoride. Silanes with halogen or sulfonate substituents are electrophilic and readily undergo nucleophilic displacement. Stannanes have the potential to act as Lewis acids when substituted by electronegative groups such as halogens. Either displacement of a halide or expansion to pentacoordinate or hexacoordinate structures is possible. In contrast to the transition metals, where there is often a change in oxidation level at the metal during the reaction, there is usually no change in oxidation level for boron, silicon, and tin compounds. The synthetically important reactions of these three groups of compounds involve transfer of a carbon substituent with one (radical equivalent) or two (carbanion equivalent) electrons to a reactive carbon center.

8. Reactions Involving Carbocations, Carbenes, and Radicals as Reactive Intermediates

Trivalent carbocations, carbanions, and radicals are the most fundamental classes of reactive intermediates. Both carbocations and carbenes have a carbon atom with six valence electrons and are therefore electron-deficient and electrophilic in character, and they have the potential for skeletal rearrangements. Radicals react through homolytic bond-breaking and bond-forming reactions involving intermediates with seven valence electrons. A common feature of these intermediates is that they are of high energy, compared to structures with completely filled valence shells. Their lifetimes are usually very short. Bond formation involving carbocations, carbenes, and radicals often occurs with low activation energies. This is particularly true for addition reactions with alkenes and other systems having bonds. These reactions replace a bond with a bond and are usually exothermic. Owing to the low barriers to bond formation, reactant conformation often plays a decisive role in the outcome of these reactions. Carbocations, carbene, and radicals frequently undergo very efficient intramolecular reactions that depend on the proximity of the reaction centers. Conversely, because of the short lifetimes of the intermediates, reactions through unfavorable conformations are unusual. Mechanistic analyses and synthetic designs that involve carbocations, carbenes, and radicals must pay particularly close attention to conformational factors.

9. Aromatic Substitution Reactions

The synthetic methods for aromaticsubstitution were among the first to be developed. These reactions provide methods for introduction of nitro groups, the halogens, sulfonic acids, and alkyl and acyl groups. The regioselectivity of these reactions depends upon the nature of the existing substituent and can be ortho, meta, or para selective. A second group of aromatic substitution reactions involves aryl diazonium ions. As for electrophilic aromatic substitution, many of the reactions of aromatic diazonium ions date to the nineteenth century. There have continued to be methodological developments for substitution reactions of diazonium intermediates. These reactions provide routes to aryl halides, cyanides, and azides, phenols, and in some cases to alkenyl derivatives. Direct nucleophilic displacement of halide and sulfonate groups from aromatic rings is difficult, although the reaction can be useful in specific cases. These reactions can occur by either addition-elimination or elimination-addition. Recently, there has been rapid development of metal ion catalysis, and old methods involving copper salts have been greatly improved. Palladium catalysts for nucleophilic substitutions have been developed and have led to better procedures. Several radical reaction have some synthetic application, including radical substitution and the $S_R N1$ reaction.

10. Oxidations

Oxidation reactions that transform a functional group to a more highly oxidized derivative by removal of hydrogen and/or addition of oxygen. There are a great many oxidation methods. As the reactions are considered, it will become evident that the material spans a broader range of mechanisms than most of the previous chapters. Owing to this range, the chapter is organized according to the functional group transformation that is accomplished. This organization facilitates comparison of the methods available for effecting a given synthetic transformation. The major sections consider the following reactions: ①oxidation of alcohols; ②addition of oxygen at double bonds; ③allylic oxidation; ④oxidative cleavage of double bonds; ⑤oxidative cleavage of other functional groups; ⑥oxidations of aldehydes and ketones; and ⑦oxidation at unfunctionalized positions. The oxidants are grouped into three classes: transition metal derivatives; oxygen, ozone, and peroxides; and other reagent.

11. Summary

A major purpose of organic synthesis at the current time is the discovery, understanding, and application of biological activity. Pharmaceutical laboratories, research foundations, and government and academic institutions throughout the world are engaged in this research. Many new compounds are synthesized to discover useful biological activity, and when activity is discovered, related compounds are synthesized to improve it. Syntheses suitable for production of drug candidate molecules are developed. Other compounds are synthesized to explore the mechanisms of biological processes. The ultimate goal is to apply this knowledge about biological activity for treatment and prevention of disease. Another major application of synthesis is in agriculture for control of insects and weeds. Organic synthesis also plays a part in the development of many consumer products, such as fragrances.

The unique power of synthesis is the ability to create new molecules and materials with valuable properties. This capacity can be used to interact with the natural world, as in the treatment of disease or the production of food, but it can also produce compounds and materials beyond the capacity of living systems. Our present world uses vast amounts of synthetic polymers, mainly derived from petroleum by synthesis. The development of nanotechnology, which envisions the application of properties at the molecular level to catalysis, energy transfer, and information management has focused attention on multimolecular arrays and systems capable of self-assembly. We can expect that in the future synthesis will bring into existence new substances with unique properties that will have impacts as profound as those resulting from syntheses of therapeutics and polymeric materials.

Words and Expressions

nucleophiles　[njuk'lifaɪlz]　n. 亲核试剂

pentacoordinate　adj. 五配位的

alkylation　[ˌælkɪ'leɪʃən]　n. 烷基取代, 烷基化

hexacoordinate　adj. 六配位的

imine　['ˌmiːn]　亚胺

carbanion　[kɑr'bænˌaˌɒn]　n. 碳酸根[基]离子

enamine　[ˌ'næmˌn]　n. 烯胺, 烯胺类

carbocations　n. 碳正离子

olefination　n. 烯化反应

carbenes　['kɑrbneˌz]　n. 碳烯, 卡宾

hydroxyalkyl　n. 羟烷基

homolytic　[həu'mɒlitik]　adj. 均裂的

alkylidene　n. 亚烷基

regioselectivity　n. 区域选择性

aldol　n. 丁间醇醛, 羟醛

aryl diazonium　芳基重氮

haloesters　n. 卤代酯

diazonium　重氮化合物

sterically　[s'terˌklˌ]　adv. 空间

cyanides　['saˌəˌnaˌdz]　n. 氰化物, 氰类化合物

platinum　['plæt(ə)nəm]　n. 铂, 白金

azides　n. 叠氮化合物

palladium　[pə'leˌdˌəm]　n. 钯, 钯金

phenols　['finɑlz]　n. 酚类, 酚类化合物

rhodium　['rəudiəm]　n. 铑

sulfonate　['sʌlfəˌneˌt]　n. 磺酸盐, 磺化

ruthenium　[ruː'θiːniəm]　n. 钌

aldehydes　['ældˌhaˌdz]　n. 乙醛, 醛类

cycloaddition　[saˌkləuə'dˌʃen]　n. 环化加成

ketones　['ketəunz]　n. 酮, 酮类

pericyclic　[pεrɪˈsɪklɪk]　*adj.* 周环的

pharmaceutical　[ˌfɑrməˈsutɪk(ə)l]　*adj.* 制药的

organometallic　[ˌɔrgənoʊməˈtælɪk]　*adj.* 有机金属的

heterogeneous　[ˌhetərəˈdʒiːniəs]　*adj.* 多相的

lanthanide　[ˈlænθənaɪd]　*n.* 镧化物,镧系元素

homogeneous　[ˌhoʊməˈdʒiːniəs]　*adj.* 单相的

tetravalent　[ˌtetrəˈveɪlənt]　*adj.* 四价的

unimolecular Rearrangement　单分子重排

pentavalent　[ˌpentəˈveɪlənt]　*adj.* 五阶的

multimolecular array　多分子阵列

indium　[ˈɪndiəm]　*n.* 铟

halide　[ˈhælaɪd]　*n.* 卤化物,卤素化合物

Notes

Victor Grignard：维克多·格里格纳德,法国化学家,因发明了格氏试剂与他的同事保罗·萨巴捷一同获得诺贝尔化学奖。

Robinson annulation：罗宾逊环合反应,是 Michael 加成反应后的某些产物,还可以进行分子内的羟醛缩合,这种通过分子内的羟醛缩合生成环己酮衍生物的合成称作罗宾逊环合。

Claisen condensation：克莱森缩合反应,是指含有 α-活泼氢的酯类在醇钠、三苯甲基钠等碱性试剂的作用下,发生缩合反应形成 β-酮酸酯类化合物的反应。

Wittigreaction：是羰基用磷叶立德变为烯烃,可用于合成各种含烯键的化合物,制备醛和酮、酯环烃、芳烃、炔类衍生物、亚氨基化合物、偶氮化合物、杂环化合物等的合成。

Exercises

1. Translate the following organic materials into Chinese.

hexanoic acid　　　3-pentenoic acid　　　2-butanone　　　2,4-hexanedione

pentanal　　　　　hexanedial　　　　　2-pentyneoctane　　　1,3-butandiyne

2. Translate the following organic materials into Chinese.

(1) At the core of this revolution is chemistry, the quintessential molecular science within which is organic chemistry, a discipline that will surely be the source of many of the major advances in chemistry, biology, medicine, materials science, and environmental science in the 21st century.

(2) Whether one seeks to understand nature or to create the new materials and medicines of the future, a key starting point is thus understanding structure and mechanism.

Lesson 4　Physical Chemistry

1. Introduction to Physical Chemistry

Physical chemistry deals with the principles of physics involved in chemical interactions. It examines: How matter behaves on a molecular and atomic level. How chemical reactions occur. Physical chemists are focused on understanding the physical properties of atoms and molecules, the way chemical reactions work, and what these properties reveal.

2. Perfect Gas

A central feature of physical chemistry is its role in building models of molecular behaviour that seek to explain observed phenomena. A prime example of this procedure is the development of a molecular model of a perfect gas in terms of a collection of molecules (or atoms) in ceaseless, essentially random motion. As well as accounting for the gas laws, this model can be used to predict the average speed at which molecules move in a gas, and its dependence on temperature. In combination with the Boltzmann distribution, the model can also be used to predict the spread of molecular speeds and its dependence on molecular mass and temperature.

The perfect gas is a starting point for the discussion of properties of all gases, and its properties are invoked throughout thermodynamics. However, actual gases, "real gases", have properties that differ from those of perfect gases, and it is necessary to be able to interpret these deviations and build the effects of molecular attractions and repulsions into the model. The discussion of real gases is another example of how initially primitive models in physical chemistry are elaborated to take into account more detailed observations.

The physical state of a sample of a substance, its physical condition, is defined by its physical properties. Two samples of the same substance that have the same physical properties are in the same state. The variables needed to specify the state of a system are the amount of substance it contains, n, the volume it occupies, V, the pressure, p, and the temperature, T.

Although in principle the state of a pure substance is specified by giving the values of n, V, p, and T, it has been established experimentally that it is sufficient to specify only three of these variables since doing so fixes the value of the fourth variable. That is, it is an experimental fact that each substance is described by an equation of state, an equation that interrelates these four variables. The general form of an equation of state is

$$p = f(T, V, n) \qquad (1)$$

This equation states that if the values of n, T, and V are known for a particular substance, then the pressure has a fixed value. Each substance is described by its own equation of state, but the explicit form of the equation is known in only a few special cases. One very important example is the equation of state of a "perfect gas", which has the form $p = nRT/V$, where R is a constant independent of the identity of the gas. The equation of state of a perfect gas was established by combining a series of empirical laws.

3. The First Law

The release of energy can be used to provide heat when a fuel burns in a furnace, to produce mechanical work when a fuel burns in an engine, and to generate electrical work when a chemical reaction pumps electrons through a circuit. Chemical reactions can be harnessed to provide heat and work, liberate energy that is unused but which gives desired products, and drive the processes of life. Thermodynamics, the study of the transformations of energy, enables the discussion of all these matters quantitatively, allowing for useful predictions.

For the purposes of thermodynamics, the universe is divided into two parts, the system and its surroundings. The system is the part of the world of interest. It may be a reaction vessel, an engine, an electrochemical cell, a biological cell, and so on. The surroundings comprise the region outside the system and are where measurements are made. The type of system depends on the characteristics of the boundary that divides it from the surroundings (Figure 1). If matter can be transferred through the boundary between the system and its surroundings the system is classified as open. If matter cannot pass through the boundary the system is classified as closed. Both open and closed systems can exchange energy with their surroundings. For example, a closed system can expand and thereby raise a weight in the surroundings; a closed system may also transfer energy to the surroundings if they are at a lower temperature. An isolated system is a closed system that has neither mechanical nor thermal contact with its surroundings.

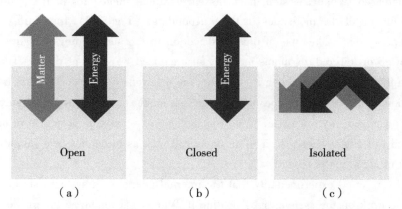

Figure 1   (a) An open system can exchange matter and energy with its surroundings. (b) A closed system can exchange energy with its surroundings, but it cannot exchange matter. (c) An isolated system canexchange neither energy nor matter with its surroundings.

The fundamental physical property in thermodynamics is work: work is done to achieve motion against an opposing force. A simple example is the process of raising a weight against the pull of gravity. A process does work if in principle it can be harnessed to raise a weight somewhere in the surroundings. An example of doing work is the expansion of a gas that pushes out a piston: the motion of the piston can in principle be used to raise a weight. Another example is a chemical reaction in a cell, which leads to an electric current that can drive a motor and be used to raise a weight. The energy of a system is its capacity to do work. When work is done on an otherwise isolated system (for instance, by compressing a gas or winding a spring), the capacity of the system to do work is increased; in other words, the energy of the system is increased. When the system does work (when the piston moves out or the spring unwinds), the energy of the system is reduced and it can do less work than before. Experiments have shown that the energy of a system may be changed by means other than work itself. When the energy of a system changes as a result of a temperature difference between the system and its surroundings the energy is said to be transferred as heat. When a heater is immersed in a beaker of water (the system), the capacity of the system to do work increases because hot water can be used to do more work than the same amount of cold water. Not all boundaries permit the transfer of energy even though there is a temperature difference between the system and its surroundings. Boundaries that do permit the transfer of energy as heat are called diathermic; those that do not are called adiabatic.

In molecular terms, heating is the transfer of energy that makes use of disorderly, apparently random, molecular motion in the surroundings. The disorderly motion of molecules is called thermal motion. The thermal motion of the molecules in the hot surroundings stimulates the molecules in the cooler system to move more vigorously and, as a result, the energy of the cooler system is increased. When a system heats its surroundings, molecules of the system stimulate the thermal motion of the molecules in the surroundings (Figure 2). In contrast, work is the transfer of energy that makes use of organized motion in the surroundings (Figure 3). When a weight is raised or lowered, its atoms move in an organized way (up or down). The atoms in a spring move in an orderly way when it is wound; the electrons in an electric current move in the same direction. When a system does work it causes atoms or electrons in its surroundings to move in an organized way. Likewise, when work is done on a system, molecules in the surroundings are used to transfer energy to it in an organized way, as the atoms in a weight are lowered or a current of electrons is passed.

It has been found experimentally that the internal energy of a system may be changed either by doing work on the system or by heating it. Whereas it might be known how the energy transfer has occurred (if a weight has been raised or lowered in the surroundings, indicating transfer of energy by doing work, or if ice has melted in the surroundings, indicating transfer of energy as heat), the system is blind to the mode employed. That is, heat and work are equivalent ways of changing the internal energy of a system. A system is like a bank: it accepts deposits in either currency (work or heat), but stores its reserves as internal energy. It is also found experimentally that if a system is isolated from its surroundings, meaning that it can exchange neither matter nor energy with its surroundings, then no change in internal energy takes place. This summary of observations is now known as the First Law of thermodynamics.

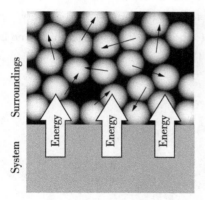

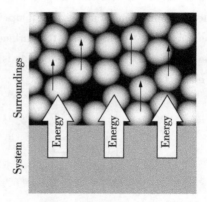

Figure 2   When energy is transferred to the surroundings as heat,
the transfer stimulates random motion of the atoms in the surroundings. Transfer of energy from the
surroundings to the system makes
use of random motion (thermal
motion) in the surroundings.

Figure 3   When a system does
work, it stimulates orderly motion in
the surroundings. For instance, the
atoms shown here may be part of
a weight that is being raised. The
ordered motion of the atoms in a
falling weight does work on the
system.

4. The Second and Third Law

Some things happen naturally, some things don' t. Some aspect of the world determines the spontaneous direction of change, the direction of change that does not require work to bring it a-bout. An important point, though, is that throughout this text "spontaneous" must be interpreted as a natural tendency which might or might not be realized in practice. Thermodynamics is silent on the rate at which a spontaneous change in fact occurs, and some spontaneous processes (such as the conversion of diamond to graphite) may be so slow that the tendency is never realized in practice whereas others (such as the expansion of a gas into a vacuum) are almost instantaneous.

The Second Law of thermodynamics expresses these conclusions more precisely and without referring to the behaviour of the molecules that are responsible for the properties of bulk matter. One statement was formulated by Kelvin: No process is possible in which the sole result is the absorption of heat from a reservoir and its complete conversion into work. Statements like this are commonly explored by thinking about an idealized device called a heat engine [Figure 4(a)]. A heat engine consists of two reservoirs, one hot (the "hot source") and one cold (the "cold sink") , connected in such a way that some of the energy flowing as heat between the two reservoirs can be converted into work. The Kelvin statement implies that it is not possible to construct a heat engine in which all the heat drawn from the hot source is completely converted into work [Figure 4(b)] : all working heat engines must have a cold sink. The Kelvin statement is a generalization of the everyday observation that a ball at rest on a surface has never been observed to leap spontaneously upwards. An upward leap of the ball would be equivalent to the spontaneous conversion of heat from the surface into the work of raising the ball. Another statement of the Second Law is due to Rudolf Clausius: Heat does not flow spontaneously from a cool body to a hotter body.

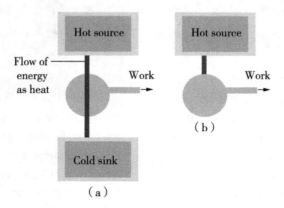

Figure 4　(a) A heat engine is a device in which energy is extracted from a hot reservoir (the hot source) as heat and then some of that energy is converted into work and the rest discarded into a cold reservoir (the cold sink) as heat. (b) The Kelvin statement of the Second Law denies the possibility of the process illustrated here, in which heat is changed completely into work, there being no other change.

To achieve the transfer of heat to a hotter body, it is necessary to do work on the system, as in a refrigerator. Although they appear somewhat different, it can be shown that the Clausius statement is logically equivalent to the Kelvin statement. One way to do so is to show that the two observations can be summarized by a single statement. First, the system and its surroundings are regarded as a single (and possibly huge) isolated system sometimes referred to as "the universe". Energy can be transferred within this isolated system between the actual system and its surroundings, but none can enter or leave it.

The entropy of an isolated system increases in the course of a spontaneous change: $\Delta S_{tot} > 0$ where S_{tot} is the total entropy of the overall isolated system. That is, if S is the entropy of the system of interest, and S_{sur} the entropy of the surroundings, then $S_{tot} = S + S_{sur}$. It is vitally important when considering applications of the Second Law to remember that it is a statement about the total entropy of the overall isolated system (the "universe"), not just about the entropy of the system of interest. The following section defines entropy and interprets it as a measure of the dispersal of energy and matter, and relates it to the empirical observations discussed so far. In summary, the First Law uses the internal energy to identify permissible changes; the Second Law uses the entropy to identify which of these permissible changes are spontaneous.

At $T=0$, all energy of thermal motion has been quenched, and in a perfect crystal all the atoms or ions are in a regular, uniform array. The localization of matter and the absence of thermal motion suggest that such materials also have zero entropy. This conclusion is consistent with the molecular interpretation of entropy because there is only one way of arranging the molecules when they are all in the ground state, which is the case at $T=0$. Thus, at $T=0$, and from $S = k\ln\Omega$, it follows that $S=0$. Entropies reported on the basis that $S_{(0)}=0$ are called Third-Law entropies (and commonly just "entropies"). When the substance is in its standard state at the temperature T, the standard (Third-Law) entropy is denoted $S^{\ominus}(T)$.

5. Chemical Equilibrium

Chemical reactions tend to move towards a dynamic equilibrium in which both reactants and products are present but have no further tendency to undergo net change. In some cases, the concentration of products in the equilibrium mixture is so much greater than that of the unchanged reactants that for all practical purposes the reaction is "complete". However, in many important cases the equilibrium mixture has significant concentrations of both reactants and products.

The spontaneity of a reaction at constant temperature and pressure can be expressed in terms of the reaction Gibbs energy: If $\Delta_r G < 0$, the forward reaction is spontaneous. If $\Delta_r G > 0$, the reverse reaction is spontaneous. If $\Delta_r G = 0$, the reaction is at equilibrium. A reaction for which $\Delta_r G < 0$ is called exergonic (from the Greek words for "work producing"). The name signifies that, because the process is spontaneous, it can be used to drive another process, such as another reaction, or used to do non-expansion work. A simple mechanical analogy is a pair of weights joined by a string: the lighter of the pair of weights will be pulled up as the heavier weight falls down. Although the lighter weight has a natural tendency to move down, its coupling to the heavier weight results in it being raised. In biological cells, the oxidation of carbohydrates acts as the heavy weight that drives other reactions forward and results in the formation of proteins from amino acids, muscle contraction, and brain activity. A reaction for which $\Delta_r G > 0$ is called endergonic (signifying "work consuming"); such a reaction can be made to occur only by doing work on it.

The increase in the number of A molecules and the corresponding decrease in the number of B molecules in the equilibrium $A(g) \rightleftharpoons 2 B(g)$ is a special case of a principle proposed by the French chemist Henri Le Chatelier, which states that:

A system at equilibrium, when subjected to a disturbance, tends to respond in a way that minimizes the effect of the disturbance.

The principle implies that, if a system at equilibrium is compressed, then the reaction will tend to adjust so as to minimize the increase in pressure. This it can do by reducing the number of particles in the gas phase, which implies a shift $A(g) \longleftarrow 2 B(g)$.

Words and Expressions

liberate [ˈlɪbəreɪt] *vt.* 解放,使自由

reservoir [ˈrezərˌvwɑ] *n.* 水库,容器

piston [ˈpɪstən] *n.* 活塞

entropy [ˈentrəpi] *n.* 熵

diathermic [ˌdaɪəˈθɜːmɪk] *adj.* 透热性的,电疗法的

exergonic [ˌeksəˈɡɒnɪk] *adj.* 放能的,放热的

adiabatic [ˌædɪəˈbætɪk] *adj.* 绝热的,隔热的

endergonic [ˌendəˈɡɒnɪk] *adj.* 吸能的,吸收能量的

spontaneous [spɒnˈteɪnɪəs] *adj.* 自发的,自然的

gibbs energy of reaction 反应吉布斯能

Notes

Boltzmann Distribution:玻尔兹曼分布。玻尔兹曼分布是热力学中的一个概念,表示在一个密闭空间系统中,温度为 *T*,系统中的不同粒子,在各种可能的量子态下的分布。

Kelvin:开尔文男爵:威廉·汤姆森,为英国的数学物理学家、工程师,是热力学第二定律的两个主要奠基人之一,也是热力学温标(绝对温标)的发明人,被称为热力学之父。

Rudolf Clausius:鲁道夫·克劳修斯,德国物理学家和数学家,热力学的主要奠基人之一。1850 年首次明确指出热力学第二定律的基本概念,于 1865 年引进了熵的概念。

Henri Le Chatelier:亨利·勒夏特列,法国化学家,勒沙特列原理是把平衡状态的某一因素加以改变之后,将使平衡状态向抵消原来因素改变的效果的方向移动。

Exercises

1. Please describe the differences between the reversible process and the irreversible process in English.

2. Please describe the first law of thermodynamics and the second law of thermodynamics, and give examples for the two laws.

Reading Material: Elements and Compounds

1. Elements

Elements are pure substances that can not be decomposed into simpler substances by ordinary chemical changes. Some common elements that are familiar to you are carbon, oxygen, aluminum, iron, copper, nitrogen, and gold. The elements are the building blocks of matter just as the numerals 0 through 9 are the building blocks for numbers. To the best of our knowledge, the elements that have been found on the earth also comprise the entire universe.

About 85% of the elements can be found in nature, usually combined with other elements in minerals and vegetable matter or in substances like water and carbon dioxide. Copper, silver, gold, and about 20 other elements can be found in highly pure forms, Sixteen elements are not found in nature; they have been produced in generally small amounts in nuclear explosions and nuclear research. They are man-made elements.

2. Pure Substances

Pure substances composed of two or more elements are called compounds. Because they contain two or more elements, compounds, unlike elements, are capable of being decomposed into simpler substances by chemical changes. The ultimate chemical decomposition of compounds produces the elements from which they are made.

The atoms of the elements in a compound are combined in whole number ratio, not in fractional parts of an atom. Atoms combined with one another to form compounds which exist as either molecule or ions. A molecule is a small, uncharged individual unit of a compound formed by the union of two or more atoms, if we subdivide a drop of water into smaller and smaller particles, we ultimately obtain a single unit of water known as a molecule of water. This water molecule consists of two hydrogen atoms and one oxygen atom bonded together. We cannot subdivide this unit further without destroying the molecule, breaking it up into its elements. Thus, a water molecule is the smallest unit of the compound water.

3. Compounds

An ionis a positive or negative electrically charged atom or group of atoms. The ions in a compound are held together in a crystalline structure by the attractive forces of their positive and negative charges. Compounds consisting of ions do not exist as molecules. Sodium chloride is an example of a non-molecular compound. Although this type of compound consists of large numbers of positive and negative ions, its formula is usually represented by the simplest ratio of the atoms in the compound. Thus, the ratio of ions in sodium chloride is one sodium ion to one chlorine ion.

Compounds exist either as molecules which consist of two or more elements bonded together or in the form of positive and negative ions held together by the attractive force of their positive and negative charges.

The compound carbon monoxide (CO) is composed of carbon and oxygen in the ratio of one atom of carbon to one atom of oxygen. Hydrogen chloride (HCl) contains a ratio of one atom of hydrogen to one atom of chlorine. Compounds may contain more than one atom of the same element. Methane ("natural gas" CH_4) is composed of carbon and hydrogen in a ratio of one carbon atom to four hydrogen atoms; ordinary table sugar(sucrose, $C_{12}H_{22}O_{11}$) contains a ratio of 12 atoms of carbon to 22 atoms of hydrogen to 11 atoms of oxygen. These atoms are held together in the compound by chemical bonds.

There are over three million known compounds, with no end in sight as to the number that can and will be prepared in the future. Each compound is unique and has characteristic physical and chemical properties. Let us consider in some detail two compounds-water and mercuric oxide. Water is a colorless, odorless, tasteless liquid that can he changed to a solid, ice, at 0 ℃ and to a gas, steam at 100 ℃. It is composed of two atoms of hydrogen and one atom of oxygen per molecule, which represents 11.2% hydrogen and 88.8% oxygen by mass. Water reacts chemically with sodium to produce hydrogen gas and sodium hydroxide, with lime to produce calcium hydroxide, and with sulfur trioxide to produce sulfuric acid. No other compound has all these exact physical and chemical properties, they are characteristic of water alone.

Mercuric oxide is a dense, orange-red powder composed of a ratio of one atom of mercury to one atom of oxygen. Its composition by mass is 92.6% mercury and 7.4% oxygen. When it is heated to temperatures greater than 360 ℃, a colorless gas, oxygen, and a silvery liquid metal, mercury, are produced. Here again are specific physical and chemical properties belonging to mercuric oxide arid to no other substance. Thus, a compound may be identified and distinguished from all other compounds by its characteristic properties.

Words and Expressions

aluminum　[ə'lumənəm]　n. 铝
nitrogen　['naɪtrədʒən]　n. 氮, 氮气
copper　['kɑːpər]　n. 铜, 铜币, 警察
silver　['sɪlvər]　n. 银
chlorine　['klɔːriːn]　n. 氯, 氯气
nuclear　['nuː kliər]　n. 核子的, 原子核的, 核能的
mercury　['mɜːrkjəri]　n. 汞, 水银
mercuric　[mər'kjʊrɪk]　adj. 汞的, 二价汞的
sodium　['soʊdiəm]　n. 钠
calcium　['kælsiəm]　n. 钙
sulfur　['sʌlfər]　n. 硫磺, 同"sulphur"
be decomposed into　被分解成
oxygen　['ɑːksɪdʒən]　n. 氧气, 氧
percent by mass　质量百分比
be composed of　由……组成
ratio of　比率
whole number ratio　整数比
positive and negative charges　正电荷和负电荷

Reading Material: Catalysis

Most of the chemical reactions in industry and biology are catalytic, and many chemists and chemical engineers work to understand and apply catalysis. Catalysis is involved at same stage of the processing of a large fraction of the goods manufactured in the United Sates; the value of these products approaches a trillion dollars annually, more than the gross national products of all but a few countries in the world. Catalysis is the key to the efficiency of chemical conversions. In an age of increasingly limited energy and raw materials and concern for the environment, it is needed more and more. The technological needs are matched by the scientific opportunities; new techniques are bringing rapid progress in the understanding of molecular details of the workings of catalyst.

Nitrogen and hydrogen flow through a tube at high temperature and pressure, but they do not react to give ammonia, although the chemical equilibrium is favorable. When particles of iron are placed in the tube, however, the gases coming in contact with them are converted rapidly into ammonia. Millions of kilograms of this chemical are synthesized every year in such tubular reactors at a cost of a few cents per kilogram. Ammonia is a raw material far nitrate fertilizers needed to feed the world' s population. Fertilizers are assimilated by the cells of plants in complex sequences of biological reactions. Further merabolic reactions take place as animals consume the plants, reconstructing their contents to provide energy and molecular building blocks for growth. All these biological reactions proceed under the direction of naturally occurring macromolecules called enzymes; without them, the processes of life could not take place.

What is an enzyme, and what does it have in common with the iron in the ammonia synthesis reactor? Each is a catalyst-a substance that accelerates a chemical reaction but is not consumed in the reaction and does not affect its equilibrium. Catalysts cause reactions to proceed faster than they would otherwise, and they can be used over and over again. Catalysts are the keys to the efficiency of almost all biochemical processes and most industrial chemical processes.

A catalyst provides a new and easier pathway for reactant molecules to be converted into product molecules. By analogy, consider instead of reactant molecules a group of climbers crossing a mountain range, breaking ground slowly over the steep crest. Alternatively, they might be guided rapidly along a less direct pathway, experiencing gentle ups and downs but circumventing the crest and crossing a low pass to reach the opposite valley in a short time. According to this rough analogy, the guidance along this new pathway-more roundabout and complicated but faster-is provided by the catalyst. The catalyst participates in the reaction by somehow combining with the reactant molecules so that they rearrange into products (and regenerate the catalyst) more rapidly than they could if the catalyst were not present.

A catalyst is by definition a substance that increases the rate of approach to equilibrium of achemical reaction without being substantially consumed in the reaction. A catalyst usually works by forming chemical bonds to one or more reactants and thereby facilitating their conversion-it does not significantly affect the reaction equilibrium.

The definition a catalyst rests on the idea of reaction rate, and therefore the subject of reaction kinetics is central, providing the quantitative framework. The qualitative chemical explanations of catalysis take the farm of reaction mechanisms. These are models of reactions accounting for the overall stoichiometry, identifying the sequence of elementary reaction steps, and (insofar as possible) explaining, in terms of chemical bond strength and geometry, the interactions of the catalyst with reactants.

Catalytic reactions take place in various phases: in solutions, within the solution like confines of micelles and the molecular-scale pockets of large enzyme molecules, within polymer gels, within the molecular-scale cages of crystalline solids such as molecular-sieve zeolites, and on the surfaces of solids. This list of phases in which catalysis occurs forms a progression from the simplest toward the most difficult to characterize in terms of exact chemical structure. A given catalytic reaction can take place in any of these phases, and the details of the mechanism will often be similar in all of them.

A full account of how a catalyst works requires a description of how reactant molecules are transported to the catalyst and of how product molecules are transported away. Because the reactants and products are often concentrated in a phase separate from that holding the catalyst, it is necessary to consider transport between phases by diffusion and convection and to determine how these transport processes affect the rates of catalytic reactions. For example, in the ammonia synthesis, the reactants H_2 and N_2 are concentrated in the gas phase flowing through the reactor, but the catalytic reaction occurs on the surface of the iron particles. Therefore, the rate of the reaction depends not only on what happens on the surface but also on haw fast the H_2 and N_2 are transported to the surface and how fast the ammonia is transported away. The reaction occurs on the surface in the steady state at just the rate of transport of the reactants to the surface and the rate of transport of the products from the surface.

Some of the essential ideas defining catalysis can he learned from preliminary consideration of an example. The reaction of ethylene and butadiene in solution to give 1,4-hexadiene is of industrial importance in the manufacture of the polymer nylon. The reaction has been suggested to proceed via the cycle of elementary steps. The species entering the cycle are H, NiL_4(a complex of Ni with unspecified ligands L), ethylene (symbolized by the double bond), and butadiene (symbolized by 1,4-butadiene). The species leaving the cycle are the hydrocarbon products, namely, three isomers of hexadiene. At this point it is not important what structures are formed from the reactants and the nickel complex. What is important is that the catalyst is formed from the reactants and the nickel complex and then converted cyclically through the various forms. The nickel is not consumed in the reaction once it has entered the cycle, but is repeatedly converted from one form to another in the cycle as the reactants are converted into products.

Catalysis always involves a cycle of reaction steps, and the catalyst is converted from one form to the next, ideally without being consumed in the overall process. The occurrence of a cyclic reaction sequence is a requirement for catalysis, and catalysis may even he defined as such an occurrence.

A catalyst for a particular reaction always provides at least one pathway for conversion of reactants that is kinetically significant. The catalyst virtually always provides this pathway as a consequence of its becoming chemically bonded to reactants.

An inhibitor glows down a catalytic reaction; a competitive inhibitor slows down the reaction by competing with the reactants in bonding to the catalyst. A very strong inhibitor, one that bonds so strongly than it virtually excludes the reactants from bonding with the catalyst, is called a poison.

A quantitative measure of how fast a catalyst works is its activity, which is usually defined as the reaction rate (or a reaction rate constant) for conversion of reactants into products. Often products are formed in addition to those that are desired, and a catalyst has an activity for each particular reaction. A ratio of these catalytic activities is referred to as selectivity, which is a measure of the catalyst's ability to direct the conversion to the desired products. For a large-scale application, the selectivity of a catalyst may be even more important than its activity. In practice, catalysts are inevitably involved in side reactions that lead to their conversion into inactive forms (those not appearing in the catalytic cycle) . Therefore, a catalyst is also chosen on the basis of its stability; the greater the stability, the lower the rate at which the catalyst loses its activity or selectivity or bath. A deactivated catalyst may be treated to bring back its activity; its regenerability in such a treatment is a measure (often not precisely defined) of how well its activity can be brought back.

Words and Expressions

account for   [əˈkaʊnt fɔː r]   （数量或比例上）占
assimilate   [əˈsɪməleɪt]   v. (使) 同化, 吸收, 消化
crystallinity   [krɪstəˈlɪnəti]   n. 结晶性
butadiene   [ˌbjuːtəˈdaɪː n]   n. 丁二烯
catalytic   [ˌkætəˈlɪtɪk]   adj. 催化(的)
crystallite   [ˈkrɪstəlaɪt]   n. 微晶, 晶粒(子), 雏晶岩
enzymatic   [enzaˈmætɪk]   adj. 酶的
equilibrate   [ˌikwəˈlaɪˌbreɪt]   v. 平衡, 使平衡
equilibrium   [ˌekwɪˈlɪbriəm]   n. 平衡, 均衡, 均势
ethylene   [ˈeθɪliːn]   n. 乙烯
facilitate   [fəˈsɪləˌteɪt]   v. 促进, 促使, 使便利
facility   [fəˈsɪləti]   n. 设施, 设备, 天赋
hexadiene   [ˈhɛksədɪən]   n. 己二烯
inactivate   [ɪnˈæktɪveɪt]   v. 使灭活, 使停止活动
macromolecular   [ˌmakroʊˈmɒlɪˌkjulər]   adj. 高分子的
Inhibitor (inhibiter)   [ɪnˈhɪbɪtə(r)]   n. 抑制剂, 约束人
molecular scale   [məˈlekjələr skeɪl]   分子尺度

polymer nylon　高分子尼龙

reactor　[riˈæktər]　*n.* 核反应堆

rearrange　[ˌriːəˈreɪndʒ]　*vt.* 重新排列,改变位置

regenerability　*n.* 再生性能

roundabout　[ˈraʊndəbaʊt]　*adj.* 迂回的,间接的

selectivity　[sɪˌlɛkˈtɪˌvəti]　*n.* 选择性,专一性

induced enzyme　[ɪnˈduːstˈenzaɪm]　诱导酶

digestive enzyme　[daɪˈdʒestɪv ˈenzaɪm]　消化酶

at the rate of　以……的速度

nitrate fertilizer　[ˈnaɪtreɪtˈfɜːrtələˌzər]　含氮肥料

Reading Material: Chemical Kinetics

Chemical kinetics and reactor design are at the heart of producing almost all industrial chemicals. It is primarily a knowledge of chemical kinetics and reactor design that distinguishes the chemical engineer from other engineers. The selection of a reaction system that operates in the safest and most efficient manner can be the key to the economic success or failure of a chemical plant. For example, if a reaction system produced a large amount of undesirable product, subsequent purification and separation of the desired product could make the entire process economically unfeasible. The chemical kinetic principles, in addition to the production of chemicals, can be applied in areas such as living systems, waste treatment and air and water pollution. Some of the examples and problems used to illustrate the principles of chemical reaction engineering are: the kinetics of nitric oxide formation and its relation to smog formation, the production of antifreeze from ethane, the manufacture of computer chips, and the application of enzyme kinetics to improve an artificial kidney.

We focus on a variety of chemical reaction engineering topics. It is concerned with the rate at which chemical reactions rake place, together with the mechanism and rate-limiting steps that control the reaction process. The sizing of chemical reactors to achieve production goals is an important segment. How materials behave within reactors, both chemically and physically, is significant to the designer of a chemical process, as is haw the data from chemical reactors should be recorded, processed, and interpreted.

We have shown that in order to calculate the time necessary to achieve a given conversion X in a batch system, or to calculate the reactor volume needed to achieve a conversion X in a flow system, we need to know the reaction rate as a function of conversion. Here we show how this functional dependence is obtained.

A homogeneous reaction is one that involves only one phase. A heterogeneous reaction involves more than one phase, and reaction usually actors at or very near the interface between the phases. An irreversible reaction is one that proceeds in only one direction and continues in that direction until the reactants are exhausted. A reversible reaction, on the other hand, can proceed in either direction, depending on the concentrations of reactants and products relative to the corresponding equilibrium concentrations. An irreversible reaction behaves as if no equilibrium condition exists. Strictly speaking, no chemical reaction is completely irreversible, but in very many reactions the equilibrium point lies so far to the right that they are treated as irreversible reactions.

In the discussion of reaction rate constant, we take as the basis of calculation a species A, which is one of the reactants that is disappearing as a result of the reactions, The limiting reactant is usually chosen as our basis for calculation. The rate of disappearance of A, $-r_A$, depends on temperature and composition. For many reactions it can be written as the product of a reaction rate constant k and a function of the concentrations (activities) of the various species involved in the reaction, i. e

$$-r_A = k(T) \cdot f(C_A C_B) \tag{1}$$

The algebraic equation that relates $-r_A$ to the species concentrations is called the kinetic expression or rate law.

The reaction rate constant k is not truly a constant, but is merely independent of the concentrations of the species involved in the reaction. The quantity k is also referred to as the specific reaction rate (constant). It is almost always strongly dependent on temperature. In gas-phase reactions, it depends on the catalyst and may be a function of total pressure. In liquid systems it can also be of function of total pressure, and in addition can depend on other parameters, such as ionic strength and choice of solvent. These other variables normally exhibit much less effect on the specific reaction rate than does temperature, so here it will he assumed that k depends only on temperature. This assumption is valid in most laboratory and industrial reactions and seems to work quite well.

It was the great Swedish chemist Arrhenius who first suggested that the temperature dependence of the specific reaction rate, k, could be correlated by an equation of the type

$$k(T) = Ae^{\frac{-E_a}{RT}} \tag{2}$$

where A = pre-exponential factor or frequency factor, E_a = activation energy, J/mol, R = gas constant = 8. 314 J/(mol $\cdot$ K) , T = absolute temperature, K.

The actuation energy E_a has been equated with a minimum energy that must be possessed by reaction molecules before the reaction will occur. From the kinetic theory of gases, the factor $e^{-E_a/RT}$ gives the fraction of the collisions between molecules that together have this minimum energy E_a. Although this might be an acceptable elementary explanation, some suggest that E_a is nothing more than an empirical parameter correlating the specific reaction rate to temperature. Other authors take exception to this interpretation. Nevertheless, postulation of the Arrhenius equation remains the greatest single step in chemical kinetics, and retains its usefulness today nearly a century later.

The activation energy is determined experimentally by carrying out the reaction at several different temperatures. After taking the natural, logarithmof Equation（2）

$$\ln k = \ln A - \frac{E_a}{R} \frac{1}{T} \tag{3}$$

It can be seen that a plot of（$\ln k$）versus（$1/T$）should be a straight line whose slope is proportional to the activation energy.

There is a rule of thumb that states that the rate of reaction doubles for every 10 ℃ increase in temperature. However, this is true only for a specific combination of activation energy and temperature. Far example, if the activation energy is 53. 6 kJ/mol, the rate will double only if the temperature is raised from 300 K to 310 K. If the activation energy in 147 kJ/mol, the rule will be valid only if the temperature is raised from 500 K to 510 K.

The larger the activation energy the more temperature-sensitive is the rate of reaction. While there are no typical values of the frequency factor and activation energy for a first order gas phase reaction, if one were forced to make a guess, values of A and E_a might be 10^3 s^{-1} and 300 kJ/mol. However, for families of reactions（e. g. , halogenation）, a number of correlations can he used to estimate the activation energy. One such correlation is the Polanyi-Semenov equation that relates activation energy to the heat of reaction. Another correlation relates activation energy to differences in hand strengths between products and reactants. While activation energy cannot be currently predicted a priori, significant research efforts are underway to calculate activation energies.

Other expressions similar to the Arrhenius equation exist. One such expression is the temperature dependence derived from transition-state theory, which takes a farm similar to Equation （2）

$$k(T) = A'T^n e^{\frac{-E_a'}{RT}} \tag{4}$$

in which $0 \leqslant n \leqslant 1$.

If equation（2）and（4）are used to describe the temperature dependence for the same reaction, it will be found that the activation energies E_a and E_a' will differ.

The dependence of the reaction rate $-r_A$ on the concentrations of the species present is almost without exception determined by experimental observation. Although the functional dependence may be postulated from theory, experiments are necessary to confirm the proposed form. One of the most common general forms of this dependence is the product of concentrations of the individual reacting species, each of which is raised to a power, e. g.

$$-r_A = k C_A^\alpha C_B^\beta \tag{5}$$

The exponents of the concentrations in Equation（5）lead to the concept of reaction order. The order of a reaction refers to which the concentrations are raised in the kinetic rate law. In Equation（5）, the reaction is α order with respect to reactant A, and β order with respect to reactant B. The overall order of the reaction, n, $n = \alpha + \beta$. In general, first- and second-order reaction are more commonly observed than zero- and third-order reactions.

Words and Expressions

kinetics　[kə'nɛtˌks]　*n.* 动力学

ethane　['iːθeˌn]　*n.* 乙烷

antifreeze　['æntifriː z]　*n.* 防冻剂, 抗凝剂

kidney　['kˌdni]　*n.* 肾, 肾脏

computerchip　[kəm'pjuːtər tʃˌps]　*n.* 电脑芯片

reversible　[rˌ'vɜːrsəbl]　*adj.* 可逆的, 可恢复原状的

artificial　[ˌɑːrtˌ'fˌʃl]　*adj.* 人造的, 人工的, 假的

pre-exponential　*n.* 前指数

segment　['segmənt]　*n.* 部分, 份, 片, 段

activation　[ˌæktə've ˌʃən]　*n.* 开动, 触发; 活化, 活性

species　['spiː ʃiː z]　*n.* 种, 物种

empirically　[ˌmˈpˌr ˌkli]　*adv.* 经验主义地

frequency　['friːkwənsi]　*n.* 频率, 重复率

collision　[kə'lˌʒn]　*n.* 碰撞, 事故, 冲突

nitric oxide　['naˌtrˌk'ɑːksaˌd]　*n.* 氧化氮, 一氧化氮

postulation　[pɒstʃə'leˌʃən]　*n.* 假定, 要求, 先决

elementary　[ˌelˌ'mentri]　*adj.* 初级的, 基本的

verificatian　[verifi'keiʃən]　*n.* 确认, 查证, 作证

interpretation　[ˌnˌtɜːrprə'teˌʃn]　*n.* 理解, 解释, 说明

slope　[sloʊp]　*n.* 斜坡, 坡地, 山坡, 斜度, 坡度

logarithm　['lɔːgərˌðəm]　*n.* 对数

halogenation　[hælədʒi'neiʃən]　*n.* 卤化, 卤化作用

power　['paʊər]　*n.* 乘方, 放大倍数

exponent　[ˌk'spoʊnənt]　*n.* 指数, 幂

Unit 2　Chemical Engineering

After completing this unit, you should be able to:

1. Describe the definition and application of chemical engineering.
2. List some core courses of chemical engineering.
3. Please tell the difference between the chemical engineers and chemists.

Lesson 1　Introduction to Chemical Engineering

Chemical engineering is concerned with transforming raw materials to final products bysubstantial chemical, biochemical or physical change of stated. It has traditionally been closely linked to the development and fortunes on the chemical and petroleum industries. These industries provide an enormous variety of products; such as synthetic rubber, gasoline and kerosene, fertilizers tailored for wheat, rice or any kind of vegetables, oxygen for firing blast furnaces or far medical uses. The discipline of chemical engineering has also always played a major part in the work of a much wider range of industries, the process industries. These include processing of food and drink, manufacture of textiles and paper, pharmaceuticals, glass and ceramics, the extraction of metals and the production of synthetic fibbers and other polymers. The chemical engineering is thus involved in the manufacturing of many goods that are bought directly by the consumer for everyday use but perhaps to a greater extent, with products that form the raw materials far manufacturing processes in a much wider range of industries.

Chemical engineering grew out of chemistry as it becomes recognized that the manufacture of chemicals required more than just a simple"scaling-up" of the chemical processes that were carried out an the laboratory bench. One of the cornerstones in its development was the concept of unit operation. Although chemical processes differ from one another, the plants in which chemicals are produced are all made up by linking together different combinations of similar processing steps or "unit operations". The underlying principles of distillation are the same whether we are distilling crude oil as a primary separation step in an oil refinery, distilling air to separate oxygen and nitrogen, or distilling the product of fermentation to produce wine; filtration of salt crystals froth a brine solution or microbial cells from a fermentation broth are all governed by the same physical processes and can all he described in a similar way. Chemical engineering thus concentrated on the general description and understanding of individual filtration, unit operations—distillation, gas absorption, crystallization, drying, extraction, leach and so on—rather than on a each individual chemical manufacturing process. So was specific description of evolved a methodology for the development of design procedures for different unit operations and for an understanding of how individual unit operations worked as a part of a complete chemical processing plant. Unit operations continue to form a central part of the discipline of chemical engineering.

One interesting outcome was the realization that the majority of operations that are involved in chemical plants are not in fact chemical at all, but are rather physical processes, the separation processes mentioned above, for example, but also heating and cooling of process streams, pumping fluids from one piece of equipment to another or transferring solid through hoppers or cyclones. Although the chemical reactor is usually at the heart of a process, other processing steps are crucial to the successful operation of the plant. The synthesis of a specific chemical product usually produces a processing stream that contains unconverted reactant and perhaps by-products. These chemical species have to be separated from one another so that unreacted material can be recycled and the required product produced at the specified purity.

A major development in understanding of chemical engineering came in the 1960s with the evolution of a"chemical engineering science" approach to the subject. By highlighting and developing a quantitative and mathematical description of the inner details of unit operations, emphasizing the links between heat transfer, mass transfer and momentum transfer (or fluid mechanics) , it put the subject on a secure and rigorous basis. The design of chemical reactors also attracted increasing attention. Chemical engineers need to understand more than just the chemical kinetics of a reaction. They need to define the extent to which the diffusion, rates of reactants and products may limit the overall reaction rate, to understand the role of heat transfer in removing or adding the heat of reaction, to design the geometry of the reactor so that the fluid flaw patterns result on effective contact between the reactants and with any catalyst that may be involved.

More recently greater emphasis has been placed on a systems approach to the study of chemical engineering where the process as a whole is studied, the interactions between different units are explored and the control scheme for, and controllability of, the process is given increasing emphasis.

As a result of these developments, chemical engineering now stands alongside civil, mechanical and electrical engineering as one of the four fundamental engineering disciplines. It is characterized by a particular approach to the solution of a wide range of problems; it deals with phenomena that aide taking place at the molecular scale and so chemical engineers must be familiar with the concepts of chemistry; it also emphasizes a system approach, looking at the overall picture as well as the innermost details of the molecular processes that are taking place. Chemical engineering is concerned with the mechanism by which reactions and separations occur and with the synthesis of the whole process. It is inherently interdisciplinary in nature, drawing on concepts from chemistry and physics, the biological sciences and other engineering disciplines, materials science and mathematics, and welding these together with its own specific areas of knowledge. And because the chemical processing industries are science-based, chemical engineering is deeply involved in basic and applied research.

Although very successful, it is probably true to say that the traditional chemical industry dies not have a good image. Itis often associated with dirt, smells and the discharge of polluting substances to the air, to rivers and to the sea. The industry is now spending vast sums of money to redress same of these problems and is coming to terms with the new challenges that it faces. It must work for a cleaner, safer environment; it meat develop more efficient ways of converting raw materials to products of high value; it must minimize the generation of waste and maximize the role of recycling; and it must develop more efficient supply and use of clean energy sources. Chemical engineers will be at the center of such developments, contributing their unique skills to these challenging problems. Perhaps their most challenging task is in relation to the impact of chemicals on the environment where they have the responsibility to act as "cradle-to-grave" guardians for chemicals. Although it is the youngest of the four major engineering disciplines, chemical engineering is now in many ways a mature discipline. But it is entering a period of rapid change and great excitement where the specific skills, knowledge and approach of chemical engineers are contributing to the solution of an ever expanding range of problems. We are sure that chemical engineering has an unlimited future.

Words and Expressions

biochemical [ˌbaɪəʊˈkemɪk(ə)l] *adj.* 生物化学的

petroleum [pəˈtrəʊliəm] *n.* 石油, 原油

synthetic [sɪnˈθetɪk] *adj.* 人造的, 合成的

ceramics [səˈræmɪks] *n.* 陶瓷, 陶瓷学, 陶瓷工艺

discipline [ˈdɪsəplɪn] *n.* 学科, 纪律, 自制力

fibber [ˈfibə] *n.* 纤维, 说谎者

pharmaceutical [ˌfɑrməˈsutɪk(ə)l] *adj.* 制药的

cornerstone [ˈkɔrnərˌstoʊn] *n.* 基石

scaling-up [skeɪlɪŋ ʌp] 按比例增加, 比例放大

distillation [ˌdistiˈleɪʃ(ə)n] *n.* 蒸馏, 精馏

concept [ˈkɒnsept] *n.* 概念, 观念, 设想, 观点

refinery [rɪˈfaɪnəri] *n.* 精制, 精炼厂

crude oil *n.* 原油

fermentation [ˌfɜrmɛnˈteɪʃən] *n.* 发酵, 发酵过程

microbial [maɪˈkrəʊbɪəl] *adj.* 细菌的, 微生物的

brine [braɪn] *n.* 盐水; *v.* 用盐水泡

gas absorption 气体吸收

broth [brɔθ] *n.* 肉汁, 肉汤, 高汤

filtration [fɪlˈtreɪʃ(ə)n] *n.* 过滤, 滤除, 滤清

crystallization　[ˌkrɪstələˈzeɪʃ(ə)n]　*n.* 结晶

unit operation　单元操作

leach　[liːtʃ]　*v.* 浸取, 浸出, 溶解出

cyclone　[ˈsaɪˌkloʊn]　*n.* 旋风, 气旋, 旋风分离器

methodology　[ˌmeθəˈdɑlədʒi]　*n.* 研究方法, 方法论

crucial　[ˈkruʃ(ə)l]　*adj.* 至关重要的, 决定性的

hopper　[ˈhɑpər]　*n.* 漏斗, V 形送料斗

unconverted　[ˌʌnkənˈvɜːtɪd]　*adj.* 无变化的

reactant　[riˈæktənt]　*n.* 反应物

by-product　*n.* 副产物

controllability　[kəntrəʊləˈbɪlɪtɪ]　*n.* 可控性, 控制能力

highlight　[ˈhaɪˌlaɪt]　*v.* 突出, 强调, 加亮

mechanical　[məˈkænɪk(ə)l]　*adj.* 机械的, 机器的

momentum　[moʊˈmentəm]　*n.* 动量, 动力, 推动力

discharge　[dɪsˈtʃɑrdʒ]　*v.* 卸料, 排放, 放出

tailor for　适合

inherently　[ɪnˈherəntlɪ]　*adv.* 固有地

civil engineering　建筑工程

interdisciplinary　[ˌɪntərˈdɪsɪplɪˌneri]　*adj.* 学科间的

electrical engineering　电子工程

process industry　过程工业, 制造(加工)工业

cradle-to-grave　终身

froth　[frɔθ]　*n.* 泡沫; *v.* 起泡沫

Exercises

1. Discuss these questions in pairs or small groups.

(1) What do industrial engineers do?

(2) What areas of expertise are especially relevant to chemical engineering?

(3) In what sorts of fields do chemical engineers work?

2. Look at the illustrations and answer the questions in pairs.

(1) What chemical processes are represented here?

(2) What products can be manufactured using these chemcial processes?

Lesson 2   Unit Operations

1. Introduction to Unit Operations

Chemical engineering has to do with industrial processes in which raw materials are changed or separated into useful products. The chemical engineer must develop, design, and engineer both the complete process and the equipment used; choose the proper raw materials; operate the plants efficiently, safely, and economically; and see to it that products meet the requirements set by the customers. Chemical engineering is both an art and a science. Whenever science helps the engineer to solve a problem, science should be used. When, as is usually the case, science does not give a complete answer, it is necessary to use experience and judgment. The professional stature of an engineer depends on skill in utilizing all sources of information to reach practical solutions to processing problems.

The range and variety of processes and industries that call for; the services of chemical engineers are both great. The field is not one that is easy to define. The processes described in standard treatises on chemical technology and the process industries give the best idea of the field of chemical engineering.

Because of the variety arid complexity of modern processes, it is not practicable to cover the entire subject matter of chemical engineering under a single head. The field is divided into convenient, but arbitrary, sectors. This text covers that portion of chemical engineering known as the unit operations.

An economical method of organizing much of the subject matter of chemical engineering is based on two facts: ①although the number of individual processes is great, each one can be broken down into a series of steps, called operations, each of which in turn appears in process after process; ②the individual operations have common techniques and are based on the same scientific principles. For example, in most processes solids and fluids must be moved, heat or other forms of energy must be transferred from one substance to another, and tasks like drying, size reduction, distillation, and evaporation must be performed. The unit operation concept is this: by studying systematically these operations themselves—operations which clearly crass industry and process lines—the treatment of all processes is unified and simplified.

The strictly chemical aspects of processing are studied in a companion area of chemical engineering called reaction kinetics. The unit operations are largely used to conduct the primarily physical steps of preparing the reactants, separating and purifying the products, recycling unconverted reactants, and controlling the energy transfer into or out of the chemical reactor.

The unit operations are as applicable to many physical processes as to chemical ones. Forexample, the process used to manufacture common salt consists of the following sequence of the unit operations: transportation of solids and liquids, transfer of heat, evaporation, crystallization, drying, and screening. No chemical reaction appears in these steps. On the other hand, the cracking of petroleum, with or without the aid of a catalyst, is a typical chemical reaction conducted on an enormous scale. Here the unit operations—transportation of fluids and solids, distillation, and various mechanical separation—are vital, and the cracking reaction could not be utilized without them. The chemical steps themselves are conducted by controlling the flow of material and energy to and from the reaction zone.

2. Classification of Unit Operations

2.1 *Fluidflow*

This concerns the principles that determine the flow or transportation of.

2.2 *Heat Transfer*

This unit operation deals with the principles that govern accumulation and transfer of heat and energy from one place to another.

2.3 *Evaporation*

This is a special case of heat transfer, which deals with the evaporation of a volatile solvent such as water from a nonvolatile solute such as salt or any other material in solution.

2.4 *Drying*

In this operation volatile liquids, usually water, are removed from solid materials.

2.5 *Distillation*

This is an operation whereby components of a liquid mixture are separated by boiling because of their differences in vapor pressure.

2.6 *Absorption*

In this process a component is removed from a gas stream by treatment with a liquid.

2.7 *Membrane Separation*

This process involves the diffusion of a solute from a liquid or gas through a semipermeable membrane barrier to another fluid.

2.8 *Liquid-liquid Extraction*

In this case a solute in a liquid solution is removed by contacting with another liquid solvent which is relatively immiscible with the solution.

2.9 *Liquid-solid Leaching*

This involves treating a finely divided solid with a liquid that dissolves out and removes a solute contained in the solid.

2.10 *Crystallization*

This concerns the removal of a solute such as a salt from a solution by precipitating the solute from the solution.

2.11 *Mechanical-physical Separations*

These involve separation of solids, liquids, or gases by mechanical means, such as filtration, settling, and size reduction, which are often classified as separate unit operations.

Many of these emit operations have certain fundamental and basic principles or mechanisms in common. For example, the mechanism of diffusion or mass transfer occurs in drying, absorption, distillation, and crystallization. Heat transfer occurs in drying, distillation, evaporation, and so on, Hence, the following classification of a more fundamental nature is often made into transfer or transport processes.

Because the unit operations are a branch of engineering, they are based on both science and experience. Theory and practice must combine to yield designs for equipment that can be fabricated, assembled, operated, and maintained. A balanced discussion of each operation requires that theory and equipment be considered together. A number of scientific principles and techniques are basic to the treatment of the unit operations. Some are elementary physical and chemical laws such as the conservation of mass and energy, physical equilibria, kinetics, and certain properties of matter.

Words and Expressions

stature [ˈstætʃər] n. 身高, 声望, 名望

treatise [ˈtritɪs] n. (专题)论文, 专著, 论述

practicable [ˈpræktɪkəb(ə)l] adj. 可行的, 行得通的

sector [ˈsektər] n. 行业, 扇形, 区域

unify [ˈjunɪfaɪ] v. 统一, 使成一体

companion [kəmˈpænjən] n. 同伴, 伴侣, 陪伴, 手册

reactant [riˈæktənt] n. 反应物

fabricate [ˈfæbrɪkeɪt] v. 制造, 编造, 捏造, 装配

assemble [əˈsemb(ə)l] v. 装配, 组装, 集合, 聚集

call for 要求

the cracking of petroleum 石油裂解

reaction kinetics 反应动力学

do with 对付, 忍耐, 与……相处

unit operation 单元操作

Exercises

1. Choose two of the terms from the box and write definitions for them that could be understood by a layperson.

evaporation, absorption, humidification, adsorption

2. In pairs, compare and discuss the types of passive voice texts that you come across in your own field of study or work.

Lesson 3　Chemical Reactors

Since all chemical processes are centered around the chemical reactor, a most important factor in determining the over—all process economy is the design of the reactor. Unlike equipment for mass—or heat—transport processes, there is no straightforward method for designing equipment to carry out a chemical reaction. This is because the design of a chemical reactor is governed primarily by the specific reaction system concerned.

In general, chemical reactors have been broadly classified in two ways, one according to the type of operation and the other according to the design features. The former classification is mainly for homogeneous reactions, and divides the reactors into batch, continuous or semi-continuous type.

1. Based on the Type of Operation

1.1 *Batch Reactor*

This type takes in all the reactants at the beginning and processes them according to a pre-determined course of reaction during which no material is fed into or removed from the reactor. Usually it is in a form of tank with or without agitation, and is used primarily in a small-scale production.

1.2 *Continuous Reactor*

Reactants are introduced and products withdrawn simultaneously in a continuous manner in this type of reactor. It may assume the shape of a tank, a tubular structure or a tower, and finds extensive applications in large-scale plants for the purpose of reducing the operating cost and facilitating control of product quality.

1.3 *Semi-continuous Reactor*

In this category belong reactors that do not fit either of the above two types. In one case some of the reactants are charged at the beginning, whereas the remaining are fed continuously as the reaction progresses. Another type is similar to a batch reactor except that one or more of the products is removed continuously.

2. Based on the Design Features

Chemical reactors have also been classified according to the design features.

2.1 *Tank Reactor*

In most instances it is equipped with some means of agitation. This type can accommodate either batch or continuous operation over wide ranges of temperatures and pressures.

2.2 *Tubular Reactor*

This type of reactor is constructed of either a single continuous tube or several tubes in parallel. The reactants enter at one end of the reactor, and the products leave from the other end, with a continuous variation in the composition of the reacting mixture in between. Heat transfer to or from the reactor may be accomplished by means of a jacket or a shell-and-tube design. The reactor tubes may be packed with catalyst pellets or inert solids.

2.3 *Tower Reactor*

A vertical cylindrical structure with a large height-to-diameter ratio characterizes this type of reactor. It may have baffles or solid packing or may be simply an empty tower, and is employed in continuous processes involving heterogeneous reactions.

2.4 *Fluidized-bed Reactor*

This is a vertical cylindrical vessel containing fine solid particles that are either catalysts or reactants.

2.5 *Bubble-phase Reactors*

Here a reagent gas is bubbled through a liquid with which it can react, because the liquid contains either a dissolved involatile catalyst or another reagent. The product may be removed from the reactor in the gas stream. Mass transfer is clearly important, and may control the rate of "reaction".

2.6 *Slurry-phase Reactor*

This type of reactor is characterized by a vertical column containing fine catalyst particles slurried with a liquid medium which may be one of the reactants.

2.7 *Trickle-bed Reactor*

In these reactors the solid catalyst is present, not as fine fluidized particles, but as a fixed bed. The reagents, which may be two partially miscible fluids, are passed either cocurrently or counter-currently through the bed.

2.8 *Moving-bed Reactor*

Here a fluid phase passes upwards through a packed bed of solid. Solid is fed to the top of the bed, moves down the column in a closely plug-flow manner, and is removed from the bottom.

Words and Expressions

batch reactor　间歇反应器
continuous reactor　连续反应器
semi-continuous reactor　半连续反应器
tank reactor　釜式反应器
tubular reactor　管式反应器
tower reactor　塔式反应器
fluidized-bed reactor　流化床反应器
bubble-phase reactor　鼓泡反应器
slurry-phase reactor　淤浆反应器
trickle-bed reactor　涓流床反应器
moving-bed reactor　移动床反应器

Exercises

Read and complete the text with the connecting words and phrases from the following word. There are two extra items.

and, because, but, due, either, since, neither, that, is, thus.

Chemical engineering is the application of chemistry—especially chemical reactions—to the process of either converting (1) <u>either</u> raw materials or chemical substances into more useful or valuable forms. Chemical engineers design, construct, and operate industrial equipment. (2) _____ they devise chemical processes that can be used to produce chemical products on a large. (3) _____, an industrial scale. (4) _____ of the diversity of the materials dealt with in industry, individual chemical engineering tasks are described in terms of unit processes. (5) _____, chemical engineers working in these areas are process engineers. (6) _____ to technological advances, the following continue to be important: the number of unit processes has increased, (7) _____ distillation, crystallization, dissolution, filtration, extraction, and polymerization.

Lesson 4   Understanding Chemistry and Chemical Engineering

1. What is Chemistry

Chemistry is a basic science whose central concerns are ①the structure and behavior of atoms (elements); ②the composition and properties of compounds; ③the reactions between substances with their accompanying energy exchange; ④the laws that unite these phenomena into a comprehensive system.

Chemistry is not an isolated discipline, for it merges into physics and biology. The origin of the term is obscure. Chemistry evolved from the medieval practice of alchemy. Its bases were laid by such men as Boyle, Lavoisier, Berzelius, Dalton and Pasteur.

In order to be eligible for entry-level positions as a chemist a bachelor's degree is required. Chemists need strong mathematical skills to make calculations and they spend a lot of time analyzing data based on their research. They primarily work indoors in laboratories and they are commonly employed in manufacturing or research. Their primary tasks involve performing experiments and assessing the data from their experiments. Their research may be used to develop new products or new ways of producing chemical compounds or they may focus on research that's intended to increase our understanding about atoms and molecules.

2. What is Chemical Engineering

Chemical Engineering is a discipline influencing numerous areas of technology. In broad terms chemical engineers are responsible for the conception and design of processes for the purpose of production, transformation and transport of materials. This activity begins with experimentation in the laboratory and is followed by implementation of the technology to full-scale production.

The large number of industries which depend on the synthesis and processing of chemicals and materials place the chemical engineer in great demand. In addition to traditional examples such as the chemical, energy and oil industries, opportunities in biotechnology, pharmaceuticals, electronic device fabrication, and environmental engineering are increasing. The unique training of the chemical engineer becomes essential in these areas whenever processes involve the chemical or physical transformation of matter. For example, chemical engineers working in the chemical industry investigate the creation of new polymeric materials with important electrical, optical or mechanical properties. This requires attention not only to the synthesis of the polymer, but also to the flow and forming processes necessary to create a final product. In biotechnology, chemical engineers have responsibilities in the design of production facilities to use microorganisms and enzymes to synthesize new drugs. Problems in environmental engineering that engage chemical engineers include the development of processes (catalytic converters, effluent treatment facilities) to minimize the release of or deactivate products harmful to the environment.

To carry out these activities, the chemical engineer requires a complete and quantitative understanding of both the engineering and scientific principles underlying these technological processes. This is reflected in the curriculum of the chemical engineering department which includes applied mathematics, material and energy balances, thermodynamics, fluid mechanics, the study of energy and mass transfer, separations technologies, chemical reaction kinetics and reactor design, and process design. These courses are built on a foundation in the sciences of chemistry, physics and biology.

Words and Expressions

academia [ˌækəˈdimiə] n. 学术界
alchemy [ˈælkəmi] n. 魔力, 炼金术
comprehensive [ˌkɑmprəˈhensɪv] adj. 综合的
deactivate [diˈæktɪˌveɪt] v. 使停止工作, 使失活
effluent [ˈefluənt] n. 流出物
fabrication [ˌfæbrɪˈkeɪʃ(ə)n] n. 制备, 制作, 加工
medieval [ˌmediˈiv(ə)l] adj. 中世纪的, 中古的
enzyme [ˈenˌzaɪm] n. 酶
microorganism [ˌmaɪkroʊˈɔrgəˌnɪzəm] n. 微生物
mentoring [ˈmentərɪŋ] n. 导师, 顾问
polymeric [ˌpɒlɪˈmerɪk] adj. 聚合态的, 聚合的
converters [kənˈvɜrtər] n. 变频器, 变换器, 换流器
fluid mechanics 流体力学

process design   工艺流程设计

Notes

1. Boyle:波义尔,英国物理学家、化学家,化学科学的开山祖师,近代化学的奠基人,1661 年波义耳所著的《怀疑派化学家》(*The Skeptical Chemist*),这一年作为近代化学的开始年代。

2. Lavoisier:拉瓦锡,法国著名化学家、生物学家,被后世尊称为"现代化学之父"。

3. Berzelius:柏济力阿斯,瑞典化学家、伯爵,现代化学命名体系的建立者。

4. Dalton:道尔顿,英国化学家、物理学家,原子理论的提出者。他所提供的关键的学说,使化学领域自那时以来有了巨大的进展。

5. Pasteur:巴斯德,法国著名的微生物学家、爱国化学家,开创了微生物生理学,其发明的巴氏消毒法至今仍被应用。

Exercises

1. Combine these three sentences into one single sentence.

Continuousor assembly-line operations are more efficient and economical than batch processes.

Continuous or assembly-line operations lend themselves to automatic control.

Chemical engineer were among the first to incorporate automatic controls into their designs.

2. Match each chemcial process with its layperson definition.

(1) crystallization   a. using a filter to separate a mixture mechanically

(2) dissolution   b. forming solid crystals from a homogenous solution

(3) distillation   c. separating substances based on differences in their
                             vapor pressures

(4) filtration   d. dissolving a substance into a liquid

(5) polymerization   e. separating the compounds of a mixture based on the
                             difference in solubility of a compound in various solvents

(6) solvent extraction   f. combiningsimple molecules to form more complex molecules
                             of highermolecular weight and with different physical
                             properties

Reading Material: Transport Phenomena

The subject of transport phenomena includes three closely related topics: fluid dynamics, heat transfer, and mass transfer. Fluid dynamics involves the transport of momentum, heat transfer deals with the transport of energy, and mass transfer is concerned with the transport of mass of various chemical species. These three transport phenomena should, at the introductory level, be studied together for the following reasons:

(1) They frequently occur simultaneously in industrial, biological, agricultural, and meteorological problems; in fact, the occurrence of any one transport process by itself is the exception rather than the rule.

(2) The basic equations that describe the three transport phenomena are closely related. The similarity of the equations under simple conditions is the basis for solving problems"by analogy".

(3) The mathematical tools needed for describing these phenomena are very similar. Although it is not the aim of this book to teach mathematics, the student will be required to review various mathematical topics as the development unfolds. Learning how to use mathematics may be a very valuable by-product of studying transport phenomena.

(4) The molecular mechanisms underlying the various transport phenomena are very closely related. All materials are made up of molecules, and the same molecular motions and interactions are responsible for viscosity, thermal conductivity, and diffusion.

A good grasp of transport phenomena is essential for understanding many processes in engineering, agriculture, meteorology, physiology, biology, analytical chemistry, materials science, pharmacy, and other areas. Transport phenomena is a well-developed and eminently useful branch of physics that pervades many areas of applied science such as biotechnology, microelectronics, nanotechnology, and polymer science, etc.

The transport of mass, momentum, energy, and angular momentumcan be described at three different levels:

(1) macroscopic level, which describe how the mass, momentum, energy, and angular momentum in the system change because of the introduction and removal of these entities via the entering and leaving streams, and because of various other inputs to the system from the surroundings. In studying an engineering or biological system it is a good idea to start with this macroscopic description in order to make a global assessment of the problem; in some instances it is only this overall view that is needed.

(2) microscopic level, which describe how the mass, momentum, energy, and angular momentum change. Its aim is to get information about velocity, temperature, pressure, and concentration profiles within the system.

(3) molecular level, which describe the mechanisms of mass, momentum, energy, and angular momentum transport in terms of molecular structure and intermolecular forces. Generally this is the realm of the theoretical physicist or physical chemist, but occasionally engineers and applied scientists have to get involved at this level. This is particularly true if the processes being studied involve complex molecules, extreme ranges of temperature and pressure, or chemically reacting systems.

It should be evident that these three levels involve different "length scales" for example, in a typical industrial problem, at the macroscopic level the dimensions of the flow systems may be of the order of centimeters or meters; the microscopic level involves what is happening in the micron to the centimeter range; and molecular level problems involve ranges of about 1 to 1 000 nanometers. There are many connections between the three levels. The transport properties that are described by molecular theory are used at the microscopic level. Furthermore, the equations developed at the microscopic level are needed in order to provide some input into problem solving at the macroscopic level. To use the macroscopic balances intelligently, it is necessary to use information about inter-phase transport that comes from the equations of change. To use the equations of change, we need the transport properties; which are described by various molecular theories. Therefore, from a teaching point of view, it seems best to start at the molecular level and work upward toward the larger systems.

There are also many connections between the three areas of momentum, energy, and mass transport. By learning how to solve problems in one area, one also learns the techniques for solving problems in another area. The similarities of the equations in the three areas mean that in many instances one can solve a problem "by analogy"—that is, by taking over a solution directly from one area and, then changing the symbols in the equations, write down the solution to a problem in another area.

Some problems involving transport phenomena in chemically reacting systems have been presented. For simplicity, the chemical kinetics expressions can be taken to be of rather idealized forms. For problems in combustion, flame propagation, and explosion phenomena more realistic descriptions of the kinetics will be needed. The same is true in biological systems, and the understanding of the functioning of the human body will require much more detailed descriptions of the interactions among chemical kinetics, catalysis, diffusion, and turbulence.

At all three levels of description—molecular, microscopic, and macroscopic—the conservation laws play a key role. Conservation laws in physics, basic laws that together determine which processes can or cannot occur in nature; each law maintains that the total value of the quantity governed by that law, e. g. , mass or energy, remains unchanged during physical processes. Conservation laws have the broadest possible application of all laws in physics and are thus considered by many scientists to be the most fundamental laws in nature.

How to study the subject of transport phenomena? Here are a few suggestions:

(1) Always read the text with pencil and paper in hand; work through the details of the mathematical developments and supply any missing steps.

(2) Whenever necessary, go back to the mathematics textbooks to brush up on calculus, differential equations, vectors, etc. This is an excellent time to review the mathematics that was learned earlier (but possibly not as carefully as it should have been).

(3) Make it a point to give a physical interpretation of key results; that is, get in the habit of relating the physical ideas to the equations.

(4) Always ask whether the results seem reasonable. If the results do not agree with intuition, it is important to find out which is incorrect.

(5) Make it a habit to check the dimensions of all results. This is one very good way of locating errors in derivations.

Words and Expressions

by analogy　照此类推

pervade　[pərˈveɪd]　v. 渗透, 弥漫, 遍及

grasp　[græsp]　n. 领会, 紧握, 控制, 紧抓

angular momentum　[ˈæŋɡjələr mouˈmentəm]　角动量

meteorology　[ˌmitiəˈrɑlədʒi]　n. 气象学

intuition　[ˌɪntuˈɪʃn]　n. (一种) 直觉, 直觉力

physiology　[ˌfɪziˈɑlədʒi]　n. 生理学, 生理机能

intermolecular force　分子间作用力

Reading Material: Introduction to Chemical Engineering Thermodynamics

1. Introduction

Separation of fluid mixtures is one of the cornerstones of chemical engineering. For rational design of a typical separation process (for example, distillation), we require thermodynamic properties of mixtures; in particular, for a system that has two or more phases at some temperature and pressure, we require the equilibrium concentrations of all components in all phases. Thermodynamics provides a tool for meeting that requirement. For many chemical products (especially commodity chemicals), the cost of separation makes a significant contribution to the total cost of production. Therefore, there is a strong economic incentive to perform separations with optimum efficiency. Thermodynamics can contribute toward that optimization.

Fifty years ago, most chemical engineering thermodynamics was based on representation of experimental data in charts, tables, and correlating equations that had little, if any, theoretical basis. Fifty years ago, most chemical engineering thermodynamics was in what we may call an empirical stage. However, the word "empirical" has several interpretations. When we represent experimental data by a table or diagram (for example, the steam tables or a Mollier diagram), we call such representation empirical. When we fit experimental ideal-gas heat capacities to, say, a quadratic function of temperature, we choose that algebraic function only because it is convenient to do so. However, when, for example, we fit vapor-pressure data as a function of temperature, we inevitably do so by expressing the logarithm of the vapor pressure as a function of the reciprocal absolute temperature. This expression is also empirical but in this case our choice of dependent and independent variables follows from a theoretical basis, viz. the Clausius-Clapeyron equation. Similarly, when for a binary mixture, we represent vapor-liquid equilibrium data with, say, the Margules equation for activity coefficients, we also call that representation empirical, although it has a theoretical foundation, viz. the Gibbs-Duhem equation. We should distinguish between "blind" empiricism, where we fit experimental data to a totally arbitrary mathematical function, and "thermodynamically-grounded" empiricism where data are expressed in terms of a mathematical function suggested by classical thermodynamics.

If now, in addition to thermodynamics, we introduce into our method of representation some more-or-less crude picture of molecular properties, for example, the van der Waals equation of state, we are still in some sense empirical but now we are in another realm of representation that we may call "phenomeno-logical" thermodynamics. The advantage of proceeding from "blind" to "thermodynamically-grounded" to "phenomenological" is not only economy in the number of adjustable parameters but also in rising ability to interpolate and (cautiously) extrapolate limited experimental data to new conditions, where experimental data are unavailable. Because "phenomenological" thermodynamics uses molecular concepts, an alternate designation is to say "molecular" thermodynamics. The most striking engineering-oriented examples of molecular thermodynamics are provided by numerous useful correlations, based on the theorem of corresponding states or on the concept of group contributions.

For engineering application, applied thermodynamics is primarily a tool for "stretching" experimental data: given some data for limited conditions, thermodynamics provides procedures for generating data at other conditions. However, thermodynamics is not magic. Without some experimental information, it cannot do anything useful. Therefore, for progress in applied thermodynamics, the role of experiment is essential: there is a pervasive need for ever more experimental results. Anyone who "does" thermodynamics is much indebted to those who work in laboratories to obtain thermodynamic properties. It is impossible here to mention even a small fraction of the vast body of new experimental results obtained over a period of fifty years. However, it is necessary here to thank the hundreds of experimentalists who have provided essential contributions to progress in chemical thermodynamics.

2. Thermodynamic Properties of Pure Fluids

For common fluids (for example, water, ammonia, light hydrocarbons, carbon dioxide, sulfur dioxide, and some freons) we have detailed thermodynamic data conveniently compiled in tables and charts; the outstanding example of such compilations is the steam table with periodic improvements and extensions. Although thermodynamic data for less common fluids are often sketchy, a substantial variety of thermodynamic properties for "normal" fluids can be estimated from corresponding-states correlations, especially those based on Pitzer's use of the acentric factor for extending and much improving classical (van der Waals) corresponding states. Here "normal" applies to fluids whose nonpolar or slightly polar (but not hydrogen-bonded) molecules are not necessarily spherical but may be quasi-elliptical. Although the Lee-Kesler tables are useful for numerous fluids, with few exceptions, correlations that are linear in the acentric factor cannot be used for strongly polar molecules or for oligomers, or other large molecules whose acentric factors exceed (roughly) 0.4.

3. Equations of State (EOS)

In the period 1950–1975, there were two major developments that persuaded chemical engineers to make more use of an EOS for fluid-phase (especially VLE) equilibria. First, in the mid-1950s, several authors suggested that successful extension of a pure-component equation of state to mixtures could be much improved by introducing one binary constant into the (somewhat arbitrary) mixing rules, that relate the constants for a mixture to its composition. For example, in the van der Waals EOS, parameter a for a mixture is written in the form

$$a(\text{mixture}) = \sum_i \sum_j z_i z_j a_{ij}$$

where i and j represent components and z is the mole fraction. When $i = j$, van der Waals constant a_{ij} is that for the pure component. When $i \neq j$, the common procedure is to calculate a_{ij} as the geometric mean corrected by ($1 - k_{ij}$) where k_{ij} is a binary parameter

$$a_{ij} = \sqrt{a_{ii} a_{jj}}(1 - k_{ij})$$

Parameter k_{ij} is obtained from some experimental data for the i-j binary.

In 1976, Peng and Robinson (PR) published their modification of the van der Waals EOS that, unlike Soave's modification (SRK), introduces a new density dependence in addition to a new temperature dependence into the RK equation. Although Soave's equation and the PR equation necessarily (by design) give good vapor pressures, the PR EOS gives better liquid densities. The PR EOS and the SRK EOS are now the most common "working horses" for calculating high-pressure VLE in the natural-gas, petroleum and petrochemical industries. For application of the PR EOS to mixtures containing polar as well as nonpolar components, a particularly useful correlation is that given by Vera and Stryjek in 1986. For mixtures where one (or more) components are well below their normal boiling points, a useful modification is that by Mathias and Copeman in 1983.

4. Activity-coefficient Models for Liquid Mixtures of Nonelectrolytes

More than one-hundred years ago, Margules proposed to correlate isothermal binary vapor-liquid equilibria (VLE) with a power series in a liquid-phase mole fraction to represent $\ln\gamma_1$, where γ_1 is the activity coefficient of component 1. The activity coefficient of component 2, γ_2 is then obtained from the Gibbs-Duhem equation without requiring additional parameters. About 15 years later, van Laar derived equations for $\ln\gamma_1$, and $\ln\gamma_2$ based on the original van der Waals equation of state. After introducing a key simplifying assumption for liquids at modest pressures (no volume change upon isothermal mixing), van Laar assumed that the isothermal entropy of mixing at constant volume is equal to that for an ideal solution. About 1930, Hildebrand and (independently) Scatchard, presented a derivation similar to that of van Laar but, instead of van der Waals constant a, they used the concept of cohesive energy density, that is, the energy required to vaporize a liquid per unit liquid volume; the square root of this cohesive energy density is the well-known solubility parameter δ. In the final Hildebrand expressions for $\ln\gamma_1$ and $\ln\gamma_2$, the square root of the cohesive energy density appears because of a geometric mean assumption similar to that used by van Laar. Because of this geometric-mean assumption, the original regular-solution theory is predictive, requiring only pure-component experimental data (vapor pressures, enthalpies of vaporization, and liquid densities).

When the modified regular-solution theory for liquid mixtures was combined with the Redlich-Kwong equation of state for vapor mixtures, it was possible to correlate a large body of VLE data for mixed hydrocarbons, including those at high pressures found in the petroleum and natural-gas industries. The resulting Chao-Seader correlation was used extensively in industry until it was replaced by other simpler methods, based on a cubic equation of state applied to all fluid phases.

Until about 1964, most chemical engineering applications of activity coefficients were based on either the Margules or the van Laar equations, although in practice, the two binary coefficients in the van Laar equation were not those based on the van der Waals equation of state, but instead, those obtained from reduction of binary VLE data.

Encouraged by Wilson's use of the local composition concept of 1964, two other models with the same concept were proposed: Renon's nonrandom two-liquid (NRTL) model of 1968, and Abrams' universal quasi-chemical (UNIQUAC) model of 1975. Although the theoretical basis of these local-composition models is not strong, subsequent to their publication, they obtained some support from molecular simulation studies. The UNIQUAC equations use only two adjustable binary parameters per binary and, because the configurational part of the excess entropy is based on Flory's expression for mixtures of noninteracting short and long-chain molecules, UNIQUAC is directly applicable to liquid mixtures that contain polymers.

Words and Expressions

rational　['ræʃnəl]　adj. 合理的, 理性的, 明智的

Margules equation　马居尔方程

empiricism　[ɪm'pɪrɪsɪzəm]　n. 实证论, 经验主义

vapor-liquid equilibrium　气液平衡

viz. [vɪz]　adv. 即, 就是

Mollier diagram　莫里尔图

theorem　['θiːərəm]　n. 尤指数学定理

Clausius-Clapeyron equation　克劳修斯-克拉佩龙方程

experimentalists　[ɪkˌsperə'mentəlˌst]　n. 实验主义者

van der Waals equation of state　范德华状态方程

arbitrary　['ɑːbɪˌtreri]　adj. 任意的, 武断的

phenomenological　[fɪˌnɒmɪnə'lɒdʒɪkl]　adj. 现象学的

geometric mean　[ˌdʒiːəˌmetrɪk 'miːn]　n. 几何平均数

acentric factor　[eˌ'sentrˌk]　*n.* 偏心因子

configurational　[kənˌfˌgjə'reˌʃən(ə)l]　*adj.* 构型

quasi-elliptical　拟(准)椭圆的,拟椭圆的

Gibbs-Duhem equation　吉布斯-杜亥姆公式

activity coefficient　活度系数

Part 3

Chemical Systematic Engineering

Unit 1　Instrumentation Automation

Answer the following questions after reading the text.
1. What components are in a test measurement system?
2. What is the function of the transducer in the measurement system?
3. What are the differences between data loggers and data analyzers?

Lesson 1　Test Measurement Instrumentation

1. Why and Where Test Measurement Instrumentation is Used

Test measurement instrumentation is used to obtain the values of the physical parameters that exist in an environment, process, or product. Examples of these physical parameters include temperature, pressure, flow, motion, force, and strain. We need to make these measurements in order to test theories and equipment, to verify that a product's specifications are met, to monitor the operation of equipment, or simply to tell us how hot it is outside. Test measurement instrumentation is used in research and development laboratories, in product development centers, in manufacturing plants, in maintenance facilities, and in commercial and household products. Research and development (R&D) laboratories are prime users of measurement instrumentation. Pure research that expands knowledge and verifies the theoretical advances in science is carried out by government labs, university labs, and the labs of large companies. The development laboratories apply the theoretical knowledge and scientific advances to improve the equipment, operation, and facilities used by society. Product development centers then design, test, and improve new consumer and industrial products based on the work of the R&D laboratories. Manufacturing plants produce the final products and use measurement instrumentation to verify that their products meet specifications. This measurement activity takes place during the production and maintenance phases of the plants' work. Measurement instrumentation is increasingly being built into commercial and household products. Automobiles are now full of sensors and displays, for example, and appliances have ever more sophisticated measurements and controls.

2. Components of a Test Measurement System

A test measurement system usually consists of several separate components for measuring some physical parameter that are joined by some means of transferring the measurement information about the parameter between the components until the information is finally presented in a form that the system operator can use. The components that are found in a typical system include the sensor, the signal conditioner, and the display. Most electrical and electronic-based systems use copper wiring as a means of transferring measurement information while other systems can use pneumatic, hydraulic, optical, or wireless transmission. In its simplest form, such as a pressure gauge, the components of a test measurement system are all contained within the same device. The sensing element (bourdon tube, bellows, diaphragm, etc.) moves in a manner that is proportional to the applied pressure. This movement is transferred to the gauge pointer through a series of gears, levers, and cams, which translate the element motion into a proportional rotational movement that drives the gauge pointer. The pointer moves to a position that is equivalent to the applied pressure, and the resulting pressure is read by the user on a circular scale located under the pointer and attached to the face of the gauge. In the case of the pressure gauge, the bourdon tube, bellows, or diaphragm is the transducer or sensor. The gears, levers, and cams form the transmission system; the pointer and dial form the display or readout. An example of a slightly more complex test measurement system is a digital pressure gauge. In this case, the sensing element, such as a diaphragm, has strain gauges attached that respond to the strain in the diaphragm material that occurs as a result of the applied pressure. Wiring connects the strain gauges to a power source and to an electronic circuit that converts the resistance change of the strain gauges into an analog voltage. Other electronics are used to linearize the voltage and provide adjustments that scale the voltage so it is proportional to the applied pressure. The analog voltage is then converted into an equivalent digital voltage, and this voltage is used to drive a digital display, which indicates the pressure in the desired units. The digital voltage can also be used to drive a liquid crystal display that mimics the traditional pressure gauge pointer and scale.

Additional electronics can provide a voltage or current output that can connect the gauge with wires to other equipment so thepressure reading can be used for remote displays, data recording, or the control of other equipment. A power source such as a battery or DC power supply must be included in the system in order to provide power for the electronics and displays. This electronics can include a microprocessor, which is used to accept the raw electronic signal and perform the data manipulation needed to provide display and output signals. In the case of a digital pressure gauge, the transducer or sensor is the diaphragm with its attached strain gauges. The sensor output and sensor power are transmitted through wires to the electronic signal conditioning circuits, and the user reads the pressure on the liquid crystal display on the gauge. An example of a complex test measurement system is the measurement of the exterior surface temperature of a space vehicle. The temperature sensor must be mounted in such a way that it is protected during the launch, space, and re-entry environments (potentially subjecting it to vibration, vacuum, radiation, space debris) while ensuring that it is still close enough to the exterior surface to make an accurate temperature measurement. The sensor's wiring is led through the space vehicle's wall in such a way that the pressure containment between the vehicle's interior and the space environment is preserved. The wiring is connected to a signal conditioning unit that provides the excitation power for the sensor and electronic circuits so the sensor's signal from the sensor can be processed into a form of signal that the measurement system can use. This signal could be an analog voltage, a digital voltage, a current, or a digital signal to be connected to a computer bus. The temperature signal could be routed to displays for the crew, to computers for data storage, and to on-board alarm systems to warn crew of abnormal conditions. The signal could also be sent by radio transmission to satellites that would then retransmit the temperature data to one or more ground stations. This would enable ground personnel to monitor the vehicle's exterior temperature.

3. Transducers

Transducers or sensors are the front end of the measurement system. They are the devices that interface directly with the equipment or environment about which we want information. They transform the physical parameter of interest (pressure, temperature, vibration, etc.) into form that can be further used and processed by other devices until the user eventually has the information he or she desires. Some transducers, such as thermocouples or piezoelectric accelerometers, have a self-generating output. In this case, the sensor element provides an electrical output through the action of the physical parameter on the sensor element. The thermocouple provides a small voltage output whenever the temperature at the junction of the two joined metal wires is different from that at the ends of the wires where the voltage is read with a voltmeter. The piezoelectric sensor element provides an electrical charge output whenever a change of force is made to the crystal sensor element. Other transducers such as resistance temperature detectors (RTDs) and strain-gauge-based transducers need a power supply in order to produce an electrical signal. In the case of the RTD, a small electrical current is passed through the resistive element so the signal-conditioning circuit can determine the resistance of the RTD and thus the temperature associated with that resistance. In the strain gauge transducer, one or more resistive strain gauges are connected in a Wheatstone bridge circuit. This Wheatstone bridge circuit operates by the application of an excitation voltage to it, which enables the resistance change of the strain gauges to be translated into an equivalent voltage change. There are thousands of transducers available commercially today, and hundreds more are introduced every year. Each transducer has a specific design that makes it more or less suitable for a given application. It is the job of the test measurement professional to select the most appropriate sensor for the application at hand. This means he or she must understand the basic operating principle of the transducer, the complete specifications of the transducer, the test environment and range of test conditions, the measurement requirements, the effect of external influences on the transducer, and the transducer's installation and calibration requirements. It is the goal of this book to give you information about a wide range of sensors so this selection process will be easier. With experience and knowledge, you will learn that certain transducer types are more suitable to certain measurement situations, simplifying the selection process.

4. Displays

Displays are the simplest form of system output device used to measure the output from the signal conditioning circuit. They indicate to the user the value of the parameter being applied to the transducer. Displays can be analog or digital meters that indicate the raw voltage from the signal conditioner, or they can be meters that scale the raw voltage so as to provide an output in units of measurement (℃ or psig). Other displays include bar graphs and alarm indicators or a combination of both. These displays can be panel mounted or bench mounted, or they can be portable hand-held units. Most of these displays rely on the user to read and note the value being measured, although some displays have some limited ability to store and retrieve data. Most displays are designed to measure static or very slowly changing data since these are the only types of signals that users can process visually. An oscilloscope, on the other hand, is an example of a display that is designed for time varying or dynamic data.

5. Data Loggers

Data loggers are standalone display systems that are used to store data from a large number of sensors. Data loggers are usually used to record data that is changing fairly slowly. A snapshot of all the data channels is taken at approximately the same time, and the data is stored along with the time at which it was taken. The scan is repeated periodically, from a few times a second to a few times an hour. The data is later retrieved from the logger's memory device and presented to the user in tabular or graphic form. Tabular outputs present the data in a table, in which the data from all or selected channels is included along with its time of acquisition. Graphical displays provide the measurement information in the form of an X-Y graph, where the x-axis shows the acquisition time and the y-axis shows the multiple data points from one or more channels. The distinction between data loggers and computer-based data acquisition systems is somewhat blurry, and today the latter have largely replaced the standalone data logger.

6. Data Acquisition System

A data acquisition system is a refinement of the data logger but with the additional flexibility that a computer-based system with software provides. The software lets the user customize the way the system acquires, stores, and displays the test data. Multiple sensors are connected to the input of the data acquisition system. The outputs of the sensors are sequentially scanned by a multiplexer, which switches each signal to a common processing circuit. This common circuit can consist of filters, analog-to-digital converters, or even a common signal conditioning circuit. Once the data is in a digital format, it can be stored, manipulated, and displayed in any number of ways. Data acquisition systems are capable of accessing hundreds of signals at speeds of over a million samples per second. They are ideally suited to measuring parameters that are changing very quickly. Data acquisition systems range in size from portable, hand-held devices that access only a few points to personal computer – based systems and even to systems controlled by mainframe computers.

7. Data Analyzers

Data analyzers are a specialized version of data acquisition systems that have additional software that processes and analyzes the sensor data. Analyzers can be designed for a specific type of measurement such as vibration analysis, or they can be general-purpose analyzers capable of manipulating data in a variety of formats. Analyzers that use the Fourier analysis technique to deal with data in the frequency and time domains are an example of the general-purpose analyzer. These analyzers can be portable and capable of accessing only a few channels, or they can be computer-based systems that access a large number of channels.

Words and Expressions

instrumentation   [ˌɪnstrəmenˈteɪʃn]   n. 仪表化

maintenance   [ˈmeɪntənəns]   n. 维护

specification   [ˌspesəfəˈkeʃən]   n. 规格

facilities   [fəˈsɪlətiz]   n. 设施(facility 的复数)

prime   [praɪm]   adj. 最初的

sophisticated [səˈfɪstɪkeɪtɪd]   adj. 复杂巧妙的, 精密的

appliance   [əˈplaɪəns]   n. 装置

pneumatic   [nuːˈmætɪk]   adj. 气动的

hydraulic   [haɪˈdrɔːlɪk]   adj. (机器)液压驱动的

gauge   [geɪdʒ]   n. 测量仪器

bellows   [ˈbeloʊz]   n. 波纹管

diaphragm   [ˈdaɪəfræm]   n. 隔膜

rotational   [roʊˈteɪʃənl]   adj. 转动的

gear   [gɪr]   n. 齿轮

analog   [ˈænəlɔːg]   adj. 模拟的

voltage   [ˈvoʊltɪdʒ]   n. 电压

mimics   [ˈmɪmɪk]   v. 模拟(mimic 的第三人称单数)

manipulation   [məˌnɪpjuˈleɪʃn]   n. 操作, 处理

microprocessor   [ˌmaɪkroʊˈpɑːsesər]   n. 微处理器

exterior   [ɪkˈstɪriər]   adj. 外部的

vibration   [vaɪˈbreɪʃn]   n. 震动

interior   [ɪnˈtɪriər]   adj. 内部的

containment   [kənˈteɪnmənt]   n. 控制, 抑制, 遏制

debris   [dəˈbriː]   n. 残骸, 碎片

thermocouple [ˈθɜːrməˌkʌpl] n. 热电偶

piezoelectric [paˌizoˌˈlektrɪk] adj. 压电的

calibration [ˌkælɪˈbreɪʃn] n. 标定, 校准

oscilloscope [əˈsɪləskoʊp] n. 示波器

dynamic [daɪˈnæmɪk] adj. 动力的

logger [ˈlɔːgər] n. 记录器

snapshot [ˈsnæpʃɑːt] n. 快照

tabular [ˈtæbjələr] adj. 列成表格的

blurry [ˈblɜːri] adj. 模糊不清的

refinement [rɪˈfaɪnmənt] n. 改进

customize [ˈkʌstəmaɪz] v. 订制, 改制

mainframe [ˈmeɪnfreɪm] n. 主机, 大型机

be transferred to   被转移到

transmit through   通过……传播

convert into   把……转化成

be processed into   被加工成……

liquid crystal display   液晶显示; 液晶显示器

Exercises

1. Please translate the following sentences into Chinese.

(1) Test measurement instrumentation is used to obtain the values of the physical parameters that exist in an environment, process, or product. Examples of these physical parameters include temperature, pressure, flow, motion, force, and strain.

(2) The sensor output and sensor power are transmitted through wires to the electronic signal conditioning circuits, and the user reads the pressure on the liquid crystal display on the gauge.

(3) Data loggers are usually used to record data that is changing fairly slowly. A snapshot of all the data channels is taken at approximately the same time, and the data is stored along with the time at which it was taken.

2. List the instrumentation during the internship and describe their effects.

Reading Material: Introduction to Industrial Automation

Industrial automation of a plant/process is the application of the process control and information systems. The world of automation has progressed at a rapid pace for the past four decades and the growth and maturity are driven by the progression in the technology, higher expectations from the users, and maturity of the industrial processing technologies. Industrial automation is a vast and diverse discipline that encompasses process, machinery, electronics, software, and information systems working together toward a common set of goals-increased production, improved quality, lower costs, and maximum flexibility.

But it' s not easy. Increased productivity can lead to lapses in quality. Keeping costs down can lower productivity. Improving quality and repeatability often impacts flexibility. It' s the ultimate balance of these four goals-productivity, quality, cost, and flexibility that allows a company to use automated manufacturing as a strategic competitive advantage in a global marketplace. This ultimate balance is difficult to achieve. However, in this case the journey is more important than the destination. Companies worldwide have achieved billions of dollars in quality and productivity improvements by automating their manufacturing processes effectively. A myriad of technical advances, faster computers, more reliable software, better networks, smarter devices, more advanced materials, and new enterprise solutions all contribute to manufacturing systems that are more powerful and agile than ever before. In short, automated manufacturing brings a whole host of advantages to the enterprise; some are incremental improvements, while others are necessary for survival. All things considered, it' s not the manufacturer who demands automation. Instead, it' s the manufacturer' s customer, and even the customer' s customer, who have forced most of the changes in how products are currently made. Consumer preferences for better products, more variety, lower costs, and "when I want it" convenience have driven the need for today' s industrial automation. Here are some of the typical expectations from the users of the automation systems.

As discussed earlier, the end users of the systems are one of the major drivers for the maturity of the automation industry and their needs are managed by the fast-growing technologies in different time zones. Here are some of the key expectations from major end users of the automation systems. The automation system has to do the process control and demonstrate the excellence in the regulatory and discrete control. The system shall provide an extensive communication and scalable architectures. In addition to the above, the users expect the systems to provide the following:

(1) Life cycle excellence from the concept to optimization. The typical systems are supplied with some cost and as a user, it is important to consider the overall cost of the system from the time the purchase is initiated to the time the system is decommissioned. This includes the cost of the system; cost of the hardware; and cost of services, parts, and support.

Single integration architecture needs to be optimum in terms of ease of integration and common database and open standards for intercommunication.

(2) Enterprise integration for the systems needs to be available for communication and data exchange with the management information systems.

(3) Cyber security protection for the systems due to the nature of the systems and their deployment in critical infrastructure. Automation systems are no more isolated from the information systems for various reasons. This ability brings vulnerability in the system and the automation system' s supplier is expected to provide the systems that are safe from cyber threats.

(4) Application integration has to be closely coupled, but tightly integrated. The systems capabilities shall be such that the integration capabilities allow the users to have flexibility to have multiple systems interconnected and function as a single system: shop floor to top floor integration or sensor to boardroom integration.

(5) Productivity and profitability through technology and services in the complete life cycle, in terms of ease of engineering, multiple locations based engineering, ease of commissioning, ease of upgrade, and migration to the newer releases.

(6) Shortening delivery time and reducing time of start-up through the use of tools and technologies. This ability clearly becomes the differentiator among the competing suppliers.

(7) SMART service capabilities in terms of better diagnostics, predictive information, remote management and diagnostics, safe handling of the abnormal situations, and also different models of business of services such as local inventory and very fast dispatch of the service engineers.

(8) Value-added services for maximization in profit, means lower product costs, scalable systems, just-in-time service, lower inventory, and technology-based services.

(9) Least cost of ownership of the control systems.

(10) Mean time to repair (MTTR) has to be minimum that can be achieved by service center at plant.

The above led to continuous research and development from the suppliers for the automation systems to develop a product that are competitive and with latest technologies and can add value to the customers by solving the main points. The following are some of the results of successful automation:

(1) Consistency: Consumers want the same experience every time they buy a product, whether it's purchased in Arizona, Argentina, Austria, or Australia.

(2) Reliability: Today's ultra efficient factories can't afford a minute of unplanned downtime, with an idle factory costing thousands of dollars per day in lost revenues.

(3) Lower costs: Especially in mature markets where product differentiation is limited, minor variations in cost can cause a customer to switch brands. Making the product as cost-effective as possible without sacrificing quality is critical to overall profitability and financial health.

(4) Flexibility: The ability to quickly change a production line on the fly (from one flavor to another, one size to another, one model to another, and the like) is critical at a time when companies strive to reduce their finished goods inventories and respond quickly to customer demands.

The earliest "automated" systems consisted of an operator turning a switch on, which would supply power to an output-typically a motor. At some point, the operator would turn the switch off, reversing the effect and removing power. These were the light-switch days of automation.

Manufacturers soon advanced to relay panels, which featured a series of switches that could be activated to bring power to a number of outputs. Relay panels functioned like switches, but allowed for more complex and precise control of operations with multiple outputs. However, banks of relay panels generated a significant amount of heat, were difficult to wire and upgrade, were prone to failure, and occupied a lot of space. These deficiencies led to the invention of the programmable controller—an electronic device that essentially replaced banks of relays—now used in several forms in millions of today's automated operations. In parallel, single-loop and analog controllers were replaced by the distributed control systems (DCSs) used in the majority of contemporary process control applications.

These new solid-state devices offered greater reliability, required less maintenance, and had a longer life than their mechanical counterparts. The programming languages that control the behavior of programmable controls and DCSs could be modified without the need to disconnect or reroute a single wire. This resulted in considerable cost savings due to reduced commissioning time and wiring expense, as well as greater flexibility in installation and troubleshooting. At the dawn of programmable controllers and DCSs, plant-floor production was isolated from the rest of the enterprise operating autonomously and out of sight from the rest of the company. Those days are almost over as companies realize that to excel they must tap into, analyze, and exploit information located on the plant floor. Whether the challenge is faster time-to-market, improved process yield, nonstop operations, or a tighter supply chain, getting the right data at the right time is essential. To achieve this, many enterprises turn to contemporary automation controls and networking architectures.

Computer-based controls for manufacturing machinery, material-handling systems, and related equipment cost-effectively generate a wealth of information about productivity, product design, quality, and delivery. Today, automation is more important than ever as companies strive to fine tune their processes and capture revenue and loyalty from consumers. This chapter will break up the major categories of hardware and software that drive industrial automation; define the various layers of automation; detail how to plan, implement, integrate, and maintain a system; and look at what technologies and practices impact manufacturers. Industrial automation is a field of engineering on application of control systems and information technologies to improve the productivity of the process, to improve the energy efficiency, to improve the safety of equipment and personnel, and to reduce the variance in the product quality and hence improve the quality.

The terminology and nomenclature of theindustrial automation systems differ based on the industry of the applications. The term for computer-integrated manufacturing (CIM) is used in the manufacturing industry context and plant wide control in a process industry context. The essential of both these terms is to interconnection of information and control systems throughout a plant in order to fully integrate the coordination and control of operations. The automation engineering spans from the sensing technologies of the physical plant variables to the networks, computing resources, display technologies, and database technologies.

Improved human operator productivity will be realized through the implementation of individual workstations, which proved the tools for decision-making as well as information that is timely, accurate, and comprehensible. Time lines of data will be assured through the interconnection of all workstations and information processing facilities with a high-speed, plant-wide LAN network and a global relational database.

The broad goal is to improve the overall process and business operations by obtaining the benefits that will come from a completely integrated plant information system. The continual growth of the linkage of the process operations data with product line, project, and business systems data will be supported. The system will make such data readily available, interactively in real time, to any employee with a need to know, at workstations scattered throughout the plant and, above all, easy to use. The resulting comprehensive plant information management system will be the key to long-range improvements to process control, product line management, plant management, and support of business strategies.

One of the major challenges in today's automation jobs is evaluating the suppliers. This challenge is more apparent in the recent days because the systems appear same across the suppliers. Here are some of the guidelines that can be considered in the selection process.

These guidelines helps to set out your organization's needs, understand how suppliers can meet them, and identify the right supplier for you. The 10 Cs are Competency, Capacity, Commitment, Control, Cash, Cost, Consistency, Culture, Clean, and Communication. Used as a checklist, the 10 Cs model can help to evaluate potential suppliers in several ways. First it helps to analyze different aspects of a supplier's business: examining all 10 elements of the checklist will give a broad understanding of the supplier's effectiveness and ability to deliver the system on time, on budget with quality while having a sustained relationship for the rest of the life cycle including engineering, installation, precommissioning, commissioning, operation, and services.

Unit 2   Chemical Process Simulation and Design

Answer the following questions after reading the text.
1. What are the importance of the chemical process simulation?
2. What kind of thermodynamics models in the chemical process simulation?
3. What is the premise of the new plant designs and control strategies?

Lesson 1   Process Simulation

Chemical process simulation has become of significant importance due to the evolution of computing tools, which have opened a wider spectrum of possibilities in the use of applications for process integration, dynamic analysis, costs evaluation, and conceptual design of reaction and separation operations. All above added to the need of performing calculations in a fast way in order to focus in the analysis of the obtained information and on other relevant aspects such as safety, green engineering, economic profitability, and many other factors that make the solutions of engineering more competitive. Process simulation is a discipline transversal to all the areas of chemical engineering. The development of many engineering projects demands simulation studies since the preliminary feasibility analysis, conceptual design, detailed design, until the process operation. For that reason, the generation of new process supported in the simulation requires the integration of concepts of chemical engineering and the breeding of innovation abilities. All that integration redounds in controllability studies and dynamic analysis, energy integration, and optimization, which aim to achieve the goals of environmental protection, process safety, and product quality.

1. Process Simulation in Chemical Engineering

Chemical process simulation aims to represent a process of chemical or physical transformation through a mathematic model that involves the calculation of mass and energy balances coupled with phase equilibrium and with transport and chemical kinetics equations. All this is made looking for the establishment (prediction) of the behavior of a process of known structure, in which some preliminary data of the equipment that constitute the process are known. The mathematical model employed in process simulation contains linear, nonlinear, and differential algebraic equations, which represent equipment or process operations, physical-chemical properties, connections between the equipment and operations and their specifications. These connections are summarized in the process flow diagram. Process flow diagrams are the language of chemical processes. Between them the state of art of an existing or hypothetical process are revealed. Thereby, the process simulators are employed for the interpretation and analysis of information contained in the process flow diagrams in order to foresee failures and evaluate the process performance. The analysis of the process is based on a mathematic model integrated by a group of equations that associate process variables such as temperature, pressure, flows, and compositions, with surface areas, geometrical configuration, set points of valves, etc. In most of the simulators the solution of the equations system is made linearly, solving each unit separately and moving forward in the system once the variables required for the calculation of the next unit are known. However, that process is useless when there are stream recycles in the system since some of the variables to calculate are required for the process initialization.

An alternative solution for that type of problems consists in taking one stream as tear stream. That means assuming the initial values of that stream to start the calculations; later on, based on the assumed information, each of the following units is solved obtaining new values for the parameters of the tear stream. Subsequently, the new values help to repeat these calculations again and again, until the difference between the initial and the calculated values fulfill a given tolerance; that point is known as convergence.

A process simulator is software used for the modeling of the behavior of a chemical process in steady state, by means of the determination of pressures, temperatures, and flows. Nowadays, the computer programs employed in process simulation have broad in the study of the dynamic behavior of processes, as well as to the control systems and their response to perturbations inherent to the operation. In the same way, software to perform equipment sizing, cost estimation, properties estimation and analysis, operability analysis and process optimization, are now available in the market; all those characteristics can be observed in the Aspen Engineering Suite. Process simulators allow:

(1) Predict the behavior of a process.

(2) Analyze in a simultaneous way different cases, changing the values of main operating variables.

(3) Optimize the operating conditions of new or existing plants.

(4) Track a chemical plant during its whole useful life, in order to foresee extensions of process improvements.

Chemical process simulation is a fundamental tool in different tasks regarding design, control, and optimization. The chemical and process engineers can develop complex calculations and evaluate different alternatives and operation scenarios. Chemical process simulators have become an essential tool in the learning of chemical engineering, and its knowledge and understanding increase the possibilities of development in the fields of conceptual and detailed engineering. The solution of simulation problems through a sequential strategy implies using tear streams over which the iterative process is made. The number of tear streams that permits the solution of a system is not necessarily the number of recycles in it. For that reason, the selection of tear streams is a careful process in which multiple variables must be taken into account, as is the number of recycles, its organization, the units involved, the sensitivity of the streams, etc.

2. Thermodynamic and Property Models

A thermodynamic model is a set of equations that permit the estimation of pure component and mixture properties. In order to represent chemical processes, their modifications, equipment or new designs, the selection of a thermodynamic model that represents accurately the physical properties of the substances interacting in such process is mandatory. Nonetheless, the importance of some properties depends on the goal of the simulation itself. For instance, if the objective is the sizing of heat exchange equipment, transport properties are vital, since these affect the equipment dynamics. Therefore, if there is a substantial error in the modeling of those properties, problems in the performance of the equipment can be evidenced after its sizing, because the real behavior of the apparatus differs from the simulated one. There are four main groups of thermodynamic models available: the ideal model, the activity coefficients models, the equations of state, and the special methods. The activity coefficients models are especially useful to describe the nonideality in the liquid phase, while the equations of state are used to calculate the nonideality in the vapor phase. However, under some conditions, the equations of state can be extrapolated to the liquid phase, and the activity coefficients models to the solid phase. Usually both methods are employed to determine the fugacity in the liquid phase.

Several aspects must be taken into account to guarantee the proper selection of the properties model. For instance, the selection of the right number of phases allows the obtainment of coherent results when interpreting the results of a model. In consequence, the selection of the adequate model and the use of the known information allow an approach to reality, necessary for the development of a good simulation project. The nonideal model does not work in most of the applications due to the fact that the molecular interactions between the components responsible of the nonidealities of the thermodynamic systems are not taken into account. On the other side, the state equations usually used are cubic equations, which combine the simplicity of the calculation with a proper approximation of the properties of most of the substances.

3. Fluid Handling Equipment

Many operations in chemical engineering involve fluids, either liquids or gases. It is therefore important to know the different options that allow transport equipment and conditioning fluid within a process. In any process plant there are pumps and piping networks gas compression systems. Fluid handling has a variety of applications, from water injection into an oil well to the steam distribution and other services in a chemical plant. Fundamental knowledge of fluid mechanics is important to understand the design of systems and equipment mentioned above, since they depend on the operation of the other processes, and similarly ensures that the flow and pressure are appropriate. It is also important to know the required power to carry out transport operations, since this information not only affects the operation from the technical point of view;

it translates directly into associated costs within the operation. Process simulation is relevant for these systems.

Process simulators studied have very similar modules that allow the calculation of equipment for fluid handling. However, each simulator has different alternatives at the time of introducing data and showing results. On one hand, Aspen Plus® allows a theoretical analysis as well as more specific data entry forms. On the other hand, Aspen HYSYS® is more geared to make a practical calculation without showing intermediate results. In both simulators it is necessary to consider the calculation module and correlations, because, as it was determined, each one counts with advantages and restrictions. In general, these methods provide a very good approximation to reality. In fact, they is now mainly used to calculate oil and gas lines, pipelines, among others. Pump and compressor modules are also used. However, these do not allow a detailed design. These provide rough estimates that in case of the equipment design are a good starting point for detailed calculations. Nonetheless, if very detailed information on the performance of the pump is required, it is possible to make a detailed rating using specialized software.

4. Heat Exchange Equipment and Heat Integration

Heat exchangers play an important role in process engineering. Their main role is adjusting the thermodynamic condition of the input stream to an operation or storage. In order to perform a thorough equipment design necessary for this operation, one must have sufficient grounds on the heat transfer mechanisms, and also, knowledge about other aspects such as construction standards, materials, and some heuristic rules that allow appropriate selection parameters to get better performance in these operations.

The simulation of heat transfer equipment has been a widely used tool in recent decades. However, in the beginning there were three independent programs: one company-oriented involved in the equipment manufacturing (design software), other involved in process engineering (evaluation program), and finally another who performed calculations in order to predict the heat exchanger performance (simulation program). These three types of programs were developed independently because the tools were not sufficiently robust, but only for specialized assessment. Subsequently based on calculation routines in design mode, because the computational complexity of such programs is substantially, higher programs were developed. And finally, the three calculation methods were integrated into a single program. A clear example of this strategy is Aspen HTFS + ® (HTFS means Heat Transfer and Fluid Flow Services) that, apart from calculations in the three modes discussed above, uses experimental information to validate results. The program focused on heat exchangers design may employ different strategies to properly design this equipment. One of them, the factorial method is to sequentially examine all possible geometries; another is to search an evolutionary search quickly routed to the most promising region, i. e. , toward the designs that are closer to the optimal, and search directed toward it. These strategies can be used taking into account the above-mentioned construction standards. Over time, the efficiency of the factorial method was improved by making some modifications to identify which calculations are simpler and thus carry out such studies first. Another very important aspect in heat exchangers design is the possibility of having parameters to optimize the cost. So, the best standard geometry that meets the requirements at the lowest cost is selected. Aspen HTFS + ® performs mechanical and hydraulic calculations that allow a more detailed design considering vibration, noise, and performance problems during operation.

In process simulators there are a variety of heat exchange modules; however, all use the same basic equations. However, there are variations in equations solving, variables involved and simplifications that can be performed to estimate heat transfer equipment. Heat exchanger modules are generally classified into process simulators according to the type of substances used in the heat exchanger and according to the phenomena that take place inside (vaporization condensation latent heat exchange only, etc.). This classification can be summarized in the following categories:

(1) Heat exchanger: both sides in a single phase, and both streams are process streams.

(2) Heater: just a phase, one process stream, and on the other side a hot utility.

(3) Cooler: one process stream; cooling is performed with water or air.

(4) Condenser: a vapor stream is condensed by using air or steam.

(5) Chiller: similar to the condenser, but in this, condensation takes place at subatmospheric pressure or temperature, so that temperature of the coolant service is other than water or air.

(6) Reboiler: exchanger is used in distillation column bottoms; you can use a service or other hotprocess stream.

In general, it is possible to define two calculation routines: short calculations and detailed calculations. The short calculation is based on estimating the amount of heat added or removed. Moreover, the detailed calculation is based on determining the area, the geometrical parameters, and heat transfer coefficients. Then, equations and principles behind the various calculation routines are illustrated.

The heat exchange modules allow calculation alternatives and are very versatile when different substances involve various processes types. Their calculation models have been among the most widely studied and still are used for calculating the different varieties, from thermodynamic exchangers to mechanically detailed designs. Aspen Plus® and Aspen HYSYS® are very useful when making preliminary equipment evaluations and its performance in the process, which can be quite valuable in situations where existing equipment will be used. On the other hand, when a rigorous heat exchanger design is required, The Aspen Exchanger Design and Rating® interface allows to perform rigorous and reliable estimates, apart from allowing application-dependent settings, where the equipment results obtained can be used for construction purposes. However, integration between Aspen Exchanger Design and Rating® and one of the process simulators can be considered in order to take advantage of the different features they provide. The presented models can become very accurate in their results, so that real equipment design from the data calculated by Aspen Exchanger Design.

5. Process Optimization in Chemical Engineering

In chemical processing units, optimization is the method that seeks to solve the problem of minimizing or maximizing an objective function that relates thevariable to optimize with the design and operating variables. The criteria for analysis of the economic objective function involve fulfilling a process criteria restrictions, conditions, design equations, and respecting the limits of the variables. Most problems in chemical engineering processes have many solutions, in some cases becoming endless. The optimization is related to the selection of an option that is best in a variety of efficient options but being the only one that comes closest to an economic optimum performance and operation. Plants operating profits are achieved by optimizing performance of the valuable products or reduction of pollutants, reducing energy consumption, improving processing flows, decreasing operation time, and minimizing plant shutdowns. To this end it is useful to identify the objective, constraints, and degrees of freedom in the process, reaching benefits such as improving the design quality, ease of troubleshooting, and a quick way to make correct decisions. However, the argument for the implementation of the optimization process is not well supported if the formulation of the optimization process has an uncertainty in the mathematical model describing the process. Although the mathematical model is a description of reality, an optimization on the mathematical model does not guarantee optimization modeling phenomenon due to the difference between the mathematical model and the actual phenomenon. In the case of chemical engineering problems, most processes and operations are well represented by mathematical models with some complexity, this leads by ensuring mathematical model optimization is going to optimize the process. Optimization can take place at any level within an organization, from a complex combination of plants, distribution facilities to each floor, units combinations, individual equipment, subsystems of a piece of equipment or smaller units. Process design and equipment specifications is usually performed by taking decisions that affect the whole life of the plant or process, which is why the right decisions are important and can be based on results optimizing both the design and daily operation of the process.

6. Dynamic Process Analysis

Dynamic analysis has been increasingly gaining importance over the last few years in the field of process design. Largely because novel process designs are more efficient, if the controllability is considered during the detailed engineering stage. By means of dynamic simulation, it is possible to monitor the behavior of the main process variables when subjected to disturbances typical of an industrial plant operation. Furthermore, the possibility of suggesting different control strategies and assess their effect on the operability makes possible to study several scenarios in a relatively short time. Generally speaking, new plant designs and control strategies must guarantee product quality, process safety, equipment protection, and compliance to environmental regulations. All these fields can be considered when developing dynamic models of existing processed or new designs.

Developing process dynamic models allows increasing the spectrum of process design enhancements during the conceptual stage. The combination of conceptual and detailed design with the establishment of the conditions that makes a process controllable and dynamically sound, becomes an alternative to generate more robust process flowsheets and unit operations. The dynamic analysis tools available in commercial process simulators make possible to accurately represent the behavior of systems in which typical dynamic situations occur, such as: dead time, disturbances, valve failure, changes in the feed characteristics, among others. These types of situations are of interest for a process designer or operator, since it allows adjusting the control mechanisms and foreseeing emergency situations in a plant.

Words and Expressions

simulator ['sɪmjuleɪtər] *n.* 模拟器

modules ['mɔdʒulz] *n.* 模块(module 的复数)

alternative [ɔːl'tɜːrnətɪv] *adj.* 可替代的,备选的

intermediate [ˌɪntər'miːdiət] *adj.* 中间的

restriction [rɪ'strɪkʃn] *n.* 限制,制约因素

spectrum ['spektrəm] *n.* 范围

transversal [træns'vɜːrsəl] *adj.* 横向的

preliminary [prɪ'lɪmɪneri] *adj.* 初步的

kinetics [kɪ'netɪks; ka'netɪks] *n.* 动力学

linear ['lɪniər] *adj.* 线性的

algebraic [ˌældʒɪ'breɪk] *adj.* 代数的

specifications [ˌspesəfə'keʃən] *n.* 规格

hypothetical [ˌhaɪpə'θetɪkl] *adj.* 假定的

variable ['veriəbl] *n.* [数]变量

geometrical [ˌdʒiːə'metrɪkl] *adj.* 几何的

configuration [kənˌfɪgjə'reɪʃn] *n.* 构造

initialization [ˌɪnɪʃələ'zeɪʃn] *n.* 初始化

tolerance ['tɑːlərəns] *n.* 公差,容许偏差

convergence [kən'vɜːrdʒəns] *n.* 收敛(性)

perturbation [ˌpɜːrtər'beɪʃn] *n.* 扰乱

scenario [sə'næriou] *n.* 场景

iterative ['ɪtəreɪtɪv,'ɪtərətɪv] *adj.* 迭代的

thermodynamic [ˌθɜːrmouda'næmɪk] *adj.* 热力学的

mandatory ['mændətɔːri] *adj.* 强制性的

extrapolate [ɪk'stræpəleɪt] *vt.* 外推

fugacity [fjuː'gæsəti] *n.* 逸度

heuristic [hju'rɪstɪk] *adj.* 探索的

Hydraulic [haɪˈdrɔːlɪk] *adj.* (与)水力学(有关)的

Chiller [ˈtʃɪlər] *n.* 制冷机

Vaporization [ˌveɪpəraˈzeɪʃn] *n.* 蒸发,蒸馏器

Utility [juːˈtɪləti] *n.* 公用工程

Coolant [ˈkuːlənt] *n.* 冷却剂

Subatmospheric [ˈsʌbætməsˈferɪk] *adj.* 低于大气压的

Rigorous [ˈrɪɡərəs] *adj.* 严格的

Versatile [ˈvɜːrsətl] *adj.* 通用的

Controllability [kənˌtrəʊləˈbɪliti] *n.* 可控性,可控制性

Algorithms [ˈælɡərɪðəm] *v.* 算法,运算法则

Robust [rəʊˈbʌst] *adj.* (系统或组织)稳固的

Operability [ɑːpərəˈbɪləti] *n.* 可操作性

be employed for   可用于

be geared to   使适应

be takeninto account   被考虑

be known as   被称为,被认为是,以……著称

Exercises

Please translate the following sentences into Chinese.

(1) Chemical process simulation aims to represent a process of chemical or physical transformation through a mathematic model that involves the calculation of mass and energy balances coupled with phase equilibrium and with transport and chemical kinetics equations.

(2) In order to represent chemical processes, their modifications, equipment or new designs, the selection of a thermodynamic model that represents accurately the physical properties of the substances interacting in such process is mandatory.

(3) The heat exchange modules allow calculation alternatives and are very versatile when different substances involve various processes types. Their calculation models have been among the most widely studied and still are used for calculating the different varieties, from thermodynamic exchangers to mechanically detailed designs.

Reading Material: Introduction to Thermodynamics in Process Simulation

At the most abstract and simple level, a process simulation model is a representation of a chemical process plant to facilitate design, analysis, or other types of studies of the behavior of that plant. Often this refers to creating a mathematical representation using computer software. All simulation models share several basic characteristics:

· Representation of the thermodynamic behavior of material.

· Representation of equipment that is encountered in a chemical process plant.

· Representation of chemical reaction connections between equipment to represent the flow of material.

The nature and the level of detail of each representation can vary greatly and the user of the simulation model must make a choice such that the results and accuracy of the model satisfy the needs of the task at hand. A model should be fit for purpose, although care should be taken not to overly simplify the model and in that way overlook possible issues in design or operation.

The benefits of simulation are not unique for gas processing plants. They apply to any chemical process plant. Simulation models bring value throughout the complete life cycle of a process. At the very early conceptual design phase of the process, the emphasis will be more on relatively simple heat and material balances. As the definition of the plant becomes more detailed, so will the model. During the conceptual phase, the results of the simulation will be used to determine approximate equipment sizes, and power and utility consumption. This will enable an estimate of the investment required for the plant and of the operating costs. The combination of these numbers with the cost of raw material and the expected market prices of the products will determine if the project is economically viable. During the front-end engineering design phase, the simulation model will provide sufficient information for a detailed design of each piece of equipment, the piping that will connect them, and the instrumentation. Dynamic simulation models will also provide insight into the controllability of the proposed design. In the detailed engineering phase and through the start-up of the plant, dynamic simulation models will provide information on the tuning of the process control system and on the validity of the proposed start-up scenarios and may help the training of the engineers and operators on the operation of the plant. Another very important aspect is the use of models to define safety equipment and to ensure that the safety system is designed to protect the plant and the people operating the plant under all incident scenarios. Finally, simulation models are invaluable during the operation of the plant for monitoring the condition of the equipment, trouble shooting, and optimizing the performance as a function of current feed product quality, market prices, and environmental factors.

At the heart of any model is the description of the behavior of the fluids that are to be processed when subject to a wide range of temperatures and pressures. The accuracy of the chosen thermodynamic model is key to the accuracy of the complete model. The selection of a thermodynamic model will depend on the nature of the components to be represented and the range of temperatures and pressures that are to be considered. Hydrocarbon molecules behave in a relatively ideal fashion at low pressure and nearambient temperature. However, most gas processing will involve highpressure operations and sometimes cryogenic temperatures. To properly model the behavior at high pressure, an equation of state (EOS) should be used. The most popular EOSs for gas processing are Peng Robinson and Soavee Redliche Kwong (SRK). When choosing a model, the complexity of the model should be taken into account. As the thermodynamic routines are called virtually all the time in a simulation model, a complex thermodynamic model will slow down the model execution. High-accuracy EOSs exist for hydrocarbon mixtures. The RefProp model from the National Institute of Standards and Technology is an example of such an EOS. This model is seldom used to model a complete flow sheet. The two reasons are that the model consumes much more central processing unit (CPU) power than an EOS like Peng Robinson and the model is less robust. The reduced robustness may lead to convergence issues when the operating conditions for which the calculations are performed fall outside of the normally expected conditions. Therefore, it is best to limit the use of complex thermodynamic models to parts of the simulation model that require that specific model to produce meaningful results. Care should be taken to avoid including these calculations in loops with many iterations. For example, the mass density of the gas may be calculated more accurately with a complex EOS. However, the gain in accuracy will often only be a few percent. For the engineering design calculations this will not make a meaningful difference. But when determining the export rate of the gas, a 1% error in mass density means a 1% impact on revenue and this translates into a considerable amount of money. However, even in this case, it is only important to calculate an accurate mass density at the export point and not throughout the whole process. In gas processing, some of the units that require special attention concerning thermodynamics are triethylene glycol (TEG) dehydration, gas sweetening, monoethylene glycol (MEG) regeneration, and cryogenic operations. The treatment of liquid water streams in general may also require special attention. More specific information can be found in the case studies section. An important element that needs to be kept in mind when selecting a thermodynamic model is that the thermodynamic model itself is only a framework represented through a set of equations. All of these equations require component-specific parameters and often also parameters that describe the interaction between components. If those parameters are not known, even the best thermodynamic model is worthless!

Once the thermodynamic model has been selected, the next choice to be made when embarking on a modeling project is whether the task at hand requires a steady-state or a dynamic simulation model. The difference between these two types of models is that a steady-state model assumes the plant is operating in a perfectly stable manner. Variation with respect to time is assumed to be zero. The vast majority of simulation models used are of the steady-state type. The reason for this popularity is that the assumption of a zero derivative with respect to time simplifies the modeling of all equipment enormously. This has a significant impact on model solution time. It also allows for a lot more flexibility in the specification of the model. In a steady-state model, it is very easy to define the model as a task to be accomplished in relatively abstract terms. For example, the process may require the pressure and temperature to be changed. A steady-state model can usually do this with a single unit operation. Although it is possible to create a similar abstract dynamic model, if the actual equipment function and geometry is not known, the dynamics of the model are arbitrary and may even lead to wrong conclusions. A dynamic simulation is typically aimed at a more realistic and more detailed model of the behavior of the process and it includes how the process behaves over time. If we consider a simple vapor-liquid separation vessel, then in a steady-state simulation the nature of the model and the inlet and outlet connections are all that is required to define the simulation. The volume in the vessel or the liquid level has no impact on the results. In a dynamic simulation, the volume and liquid level are key elements of the simulation. The aim of the simulation could, for example, be to study the level control of the vessel or the pressure transients in the vessel. The additional effort required to create a dynamic simulation model is the second reason why steady-state models are used more than dynamic models. For any modeling effort that is aimed at main equipment design, the initial modeling choice will always be a steady-state model. A dynamic model may be used at a later stage to validate the process behavior given the design choices resulting from the steady-state model combined with specific circumstances. If the aim of a simulation is related to process control or to safety, a dynamic simulation is used as the phenomena to be studied are always varying in time.

Shortcut models tend to describe the process in more abstract terms; the unit operations used do not necessarily directly represent actual plant equipment but sometimes a combination of several equipment items. Rigorous models will model each equipment item individually. There is no single definition of rigorous. Rigor can be introduced by increasing the number of unit operations used to model the system and it can also be introduced by changing the methods used inside of a unit operations model to represent the function of the equipment. As an illustration, consider a small part of the plant that has a pump to increase the pressure of a liquid and a heat exchanger that reduces the temperature of the liquid. A shortcut model could represent this as a single unit operation. The model would provide limited information for the design of the pump or the heat exchanger, but it does bring the liquid to the expected state to feed it in the next unit operation. A first step toward a more rigorous model would be to model this as a pump and a cooler. A next step would be to use a heat exchanger model that also represents the cooling fluid. The model could be further detailed by including geometry information for the heat exchanger or pump performance curves. For a dynamic model some the control valves should be added such that the behavior of controllers can be checked. The model detail can be extended even more by also including the spare pump that will be installed, the block valve that will be included, the minimum flow line on the pump, and the bypass circuit on the exchanger. Additional details like the inertia of the pump can be added for studying the start-up of the pump. The choice of the level of detail used depends on the purpose of the model. Clearly, the shortcut model will not be of any help investigating the behavior of the system when it switches to the spare pump in case of failure of the normally operating pump. However, if the currently required information is pump power and utilities consumption, a model slightly more rigorous than the shortcut model is sufficient and additional modeling of the spares and bypasses would be a waste of effort. Another typical example would be the case where a larger plant includes a TEG dehydration unit and it is already known that this will be subcontracted to a specialist vendor. The main model can represent the whole TEG dehydration unit as a single simplified block that produces a dehydrated gas with the water content specification demanded from the specialist vendor.

Lumped parameter and distributed models come into play when considering rigorous equipment models. It is a continuation of the sort of choices that are made when choosing between shortcut and rigorous models at the level of a single piece of equipment. A totally lumped parameter model will consider the content of one equipment to be homogeneous; it will not consider radial or axial gradients in the fluid properties. A fully distributed model will consider the variation of fluid properties and interaction with its neighboring elements in all three dimensions and over time. Computational fluid dynamics is an example of a method that can be a fully distributed model. As with shortcut and rigorous process models, there is a continuum of variations between the two extremes. This can also best be illustrated with some examples. A totally lumped model could be a phase separation vessel model that only predicts phase separation and a uniform heat input from the environment. This would be a zero dimensional model. A first (small) step toward a distributed model is to calculate the heat input for the part of the vessel that contains the vapor phase and the part that contains the liquid phase separately. The next step could be to drop the assumption of phase equilibrium and to calculate phase compositions and temperatures using the rate of exchange of material and heat between the phases. A practical example of this is the selection of the model used for determining the result of depressuring a vessel over time. If the calculated temperatures approach the limit where a different material of construction should be used, it would be prudent to increase the rigor of the model and hence to move to a more distributed model. Next we could consider the problem of the behavior of a large liquefied natural gas tank. A known phenomenon is roll over. The model that may have been satisfactory to describe vessel depressuring would not capture the rollover phenomenon. The model would also consider that in the liquid phase there can be multiple layers with different compositions and properties. This would lead to a one-dimensional model. If the effect of the pitch and roll of the ship are lumped together in a single parameter, our previous one-dimesional model would be expanded to be a two dimensional model. If the effects of pitch and roll are quantified separately, then the resulting model would be a three-dimensional model.

The type of model selected should always depend on the purpose of the model and ultimately it will be the economics that determine the choice. If it takes a 1 million dollar study to quantify the additional effect of the sun heating a vessel in a nonuniform way and the potential gain from the results of the study is 25 000 dollars per year, it is highly unlikely that such a study would be done.

As a bespoke model will typically be more expensive than an off-the-shelf model, a bespoke model would only be selected if off-the-shelf models cannot provide the required functionality of the model. An intermediate solution should also be investigated. Rather than creating a model from scratch, it may be more economical to use an off-the-shelf tool and construct the required model from elements provided by the off-the-shelf tool. Another element may come into play here. If the new model will likely be useful for several applications, then the added cost may be offset by the cost one would otherwise incur by reimplementing a particular solution over and over again in an off-the-shelf tool. Another important factor may be that once the bespoke model has been validated, it does not require revalidation for the next applications.

Lesson 2　Process Analysis and Integration

1. Process Optimization

Optimization is the process of improving an existing situation, device, or system such as a chemical process. In optimization, various terms are used to simplify discussions and explanations. These are defined below. Decision variables are those independent variables over which the engineer has some control. These can be continuous variables such as temperature, or discrete (integer) variables such as number of stages in a column. Decision variables are also called design variables. An objective function is a mathematical function that, for the best values of the decision variables, reaches a minimum (or a maximum). Thus, the objective function is the measure of value or goodness for the optimization problem. If it is a profit, one searches for its maximum. If it is a cost, one searches for its minimum. There may be more than one objective function for a given optimization problem. Constraints are limitations on the values of decision variables. These may be linear or nonlinear, and they may involve more than one decision variable. When a constraint is written as an equality involving two or more decision variables, it is called an equality constraint. For example, a reaction may require a specific oxygen concentration in the combined feed to the reactor. The mole balance on the oxygen in the reactor feed is an equality constraint. When a constraint is written as an inequality involving one or more decision variables, it is called an inequality constraint. For example, the catalyst may operate effectively only below 400 ℃, or below 20 MPa. An equality constraint effectively reduces the dimensionality (the number of truly independent decision variables) of the optimization problem. Inequality constraints reduce (and often bound) the search space of the decision variables. A global optimum is a point at which the objective function is the best for all allowable values of the decision variables. There is no better acceptable solution. A local optimum is a point from which no small, allowable change in decision variables in any direction will improve the objective function. Certain classes of optimization problems are given names. If the objective function is linear in all decision variables and all constraints are linear, the optimization method is called linear programming. Linear programming problems are inherently easier than other problems and are generally solved with specialized algorithms. All other optimization problems are called nonlinear programming. If the objective function is second order in the decision variables and the constraints are linear, the nonlinear optimization method is called quadratic programming.

For optimization problems involving both discrete and continuous decision variables, the adjective mixed-integer is used. Although these designations are used in the optimization literature, this chapter mainly deals with the general class of problems known as MINLP, mixed-integer nonlinear programming. The use of linear and quadratic programming is limited to a relatively small class of problems. Some examples in which these methods are used include the optimal blending of gasoline and diesel products and the optimal use of manufacturing machinery. Unfortunately, many of the constraints in chemical processes are not linear, and the variables are often a mixture of continuous and integer. A simple example of a chemical engineering problem is the evaluation of the optimal heat exchanger to use in order to heat a stream from 3 ℃ to 160 ℃. This simple problem includes continuous variables, such as the area of the heat exchanger and the temperature of the process stream, and integer variables, such as whether to use low-, medium-, or high-pressure steam as the heating medium. Moreover, there are constraints such as the materials of construction that depend, nonlinearly, on factors such as the pressure, temperature, and composition of the process and utility streams. Clearly, most chemical process problems are quite involved, and care must be taken to consider all the constraints when evaluating them.

2. Pinch Technology

Whenever the design of a system is considered, limits exist that constrain the design. These limits often manifest themselves as mechanical constraints. For example, a distillation of two components that requires 400 equilibrium stages and a tower with a diameter of 20 m would not be attempted, because the construction of such a tower would be virtually impossible with current manufacturing techniques. A combination of towers in series and parallel might be considered but would be very expensive. These mechanical limitations are often (but not always) a result of a constraint in the process design. The example of the distillation column given above is a result of the difficulty in separating two components with similar volatility. When designing heat exchangers and other unit operations, limitations imposed by the first and second laws of thermodynamics constrain what can be done with such equipment. For example, in a heat exchanger, a close approach between hot and cold streams requires a large heat transfer area. Likewise, in a distillation column, as the reflux ratio approaches the minimum value for a given separation, the number of equilibrium stages becomes very large. Whenever the driving forces for heat or mass exchange are small, the equipment needed for transfer becomes large and it is said that the design has a pinch. When considering systems of many heat- or mass-exchange devices (called exchanger networks) , there will exist somewhere in the system a point where the driving force for energy or mass exchange is a minimum. This represents a pinch or pinch point. The successful design of these networks involves defining where the pinch exists and using the information at the pinch point to design the whole network. This design process is designed as pinch technology. The concepts of pinch technology can be applied to a wide variety of problems in heat and mass transfer. As with other problems encountered in this text, both design and performance cases can be considered. The focus is on the implementation of pinch technology to new processes for both heat-exchanger networks (HENs) and mass-exchanger networks (MENs) . Retrofitting an existing process for heat or mass conservation is an important but more complicated problem. The optimization of such a retrofit must consider the reuse of existing equipment, and this involves extensive research into the conditions that exist within the process, the suitability of materials of construction to new services, and a host of other issues. By considering the design of a heat- (or mass-) exchange network for existing systems, the solution that minimizes the use of utility streams can be identified, and this can be used to guide the retrofit to this minimum utility usage goal. The approach followed in the remainder consists of establishing an algorithm for designing a heat- (mass-) exchanger network that consumes the minimum amount of utilities and requires the minimum number of exchangers (MUMNE) . Although this network may not be optimal in an economic sense, it does represent a feasible solution and will often be close to the optimum.

3. Heat Integration and Network Design

Even in a preliminary design, some form of heat integration is usually employed. Heat integration has been around in one form or another ever since thermal engineering came into being. Its early use in the process industries was most apparent in the crude preheat trains used in oil refining. In refineries, the thermal energy contained in the various product streams is used to preheat the crude prior to final heating in the fired heater, upstream of the atmospheric column. Because refineries often process large quantities of oil, product streams are also large and contain huge amounts of thermal energy. Even when energy costs were very low, the integration of the energy contained in process streams made good economic sense and was therefore practiced routinely. The growing importance of heat integration in the chemical process industries (CPI) can be traced to the large increase in the cost of fuel/energy starting in the early 1970s. As the PFD evolves, the need to heat and cool process streams becomes apparent. For example, feed usually enters a process from a storage vessel that is maintained at ambient temperature. If the feed is to be reacted at an elevated temperature, then it must be heated. Likewise, after the reaction has taken place, the reactor effluent stream must be purified, which usually requires cooling the stream, and possibly condensing it, prior to separating it. Thus, energy must first be added and then removed from the process. The concept of heat integration, in its simplest form, is to find matches between heat additions and heat removals within the process. In this way, the total utilities that are used to perform these energy transfers can be minimized, or rather optimized.

Words and Expressions

variables   ['verɪəbl]   n. 变量
integer   ['ɪntɪdʒər]   n. 整数, 统一体
Constraints   [kən'streɪnts]   n. 限制, 约束条件
nonlinear   [ˌnɑːn'lɪniər]   adj. 非线性的

concentration [ˌkɑːnsnˈtreɪʃn] *n.* 含量, 浓度

dimensionality [dɪˌmenʃəˈnælətɪ] *n.* 维度

bound [baʊnd] *v.* 绑定(bind 的过去式和过去分词)

optimum [ˈɑːptɪməm] *n.* 最佳条件

algorithms [ˈælɡərˌðəmz] *n.* 算法(algorithm 的复数)

quadratic [kwɑːˈdrætɪk] *adj.* 二次的

blending [ˈblendɪŋ] *n.* 混合

gasoline [ˈɡæsəliːn] *n.* 汽油

diesel [ˈdiːzl] *n.* 柴油

Pinch [pɪntʃ] *v.* 夹点

manifest [ˈmænɪfest] *v.* 显示, 表明

distillation [ˌdɪstɪˈleɪʃn] *n.* 精馏, 蒸馏

virtually [ˈvɜːrtʃuəli] *adv.* 模拟地

volatility [ˌvɑːləˈtɪləti] *n.* 挥发性

Imposed [ɪmˈpəʊzd] *v.* 把……强加于

thermodynamics [ˌθɜːrmoʊdaɪˈnæmɪks] *n.* 热力学

reflux [ˈriːflʌks] *n.* 回流

implementation [ˌɪmplɪmenˈteɪʃn] *n.* 实施, 执行

retrofit [ˈretroʊfɪt] *vt.* 改进

Integration [ˌɪntɪˈɡreɪʃn] *n.* 集成

preliminary [prɪˈlɪmɪneri] *adj.* 初步的

preheat [ˌpriːˈhiːt] *vt.* 预先加热

atmospheric [ˌætməsˈfɪrɪk] *adj.* 常压的

formalization [ˌfɔːrmələˈzeɪʃn] *n.* 形式化

vessel [ˈvesl] *n.* 容器

effluent [ˈefluənt] *adj.* 流出的, 发出的

be transferred to 被转移到……

mixed-integer 混合整数

be designed as 被设计成……

be traced to 追溯到……

Exercises

1. Please translate the following sentences into Chinese.

(1) An objective function is a mathematical function that, for the best values of the decision variables, reaches a minimum (or a maximum). Thus, the objective function is the measure of value or goodness for the optimization problem. If it is a profit, one searches for its maximum. If it is a cost, one searches for its minimum. There may be more than one objective function for a given optimization problem.

(2) For example, in a heat exchanger, a close approach between hot and cold streams requires a large heat transfer area. Likewise, in a distillation column, as the reflux ratio approaches the minimum value for a given separation, the number of equilibrium stages becomes very large.

2. Answer the following questions according to the text.

(1) What is the aim of the process optimization?

(2) What is the pinch technology?

(3) Is it true that a close approach between hot and cold streams requires a large heat transfer area?

(4) What does HENs stand for?

Reading Material: Exchanger Design in Different Simulators

Both simulators, Aspen Plus® and Aspen HYSYS® count with a large variety of models to calculate different heat exchange equipment. Some of these modules allow rigorously evaluating shell and tube heat exchangers. In both programs, all geometry parameters without detailed mechanical information can be specified; this allows calculating pressure drop, film coefficients, among other important variables for chemical and process engineering. Despite that those modules cannot evaluate cost, vibration, and equipment real dimensions, the results can be exported to Aspen Exchanger Design and Rating®, where the calculation can be refined.

Another alternative available is the opposite process, i. e., enter the information directly into the Aspen Exchanger Design and Rating® interface and after the calculation, use generated file for inclusion in Aspen Plus® or Aspen HYSYS® to use information previously generated for a specialized software. This process is useful when a complete process is simulated with a medium or high detail level. Additionally, when there is a change in the main process simulation, the exchanger calculation is performed externally in Aspen Exchanger Design and Rating® and the corresponding simulation results are imported. There are many differences when entering the data and how to calculate the equipment between Aspen Plus® and Aspen HYSYS®. In Aspen HYSYS® you can specify the shell and heads, while in Aspen Plus® can only specify the TEMA type of the shell. Similarly, in Aspen Plus® the tubes material can be specified based on information of different material properties commonly used for this function. However, in Aspen HYSYS® is not allowed to specify the material such as, and instead one should provide the material properties.

In terms of results between simulators, there are some differences. In Aspen Plus®, a real area very close to the value obtained in the example exchanger was obtained. In Aspen HYSYS®, the results have small variations, especially in the water flow. This difference is due to the way both simulators calculate; Aspen HYSYS®, in order to include the n-propanol restriction, immediately calculates the necessary water flow, depending on the amount of transferable energy, and does not give the chance to define the final temperature. When looking at the results, there is a large decrease in water flow and a corresponding increase in temperature. However, both simulators allow achieving the required specification for the n-propanol outlet. The simulators differ in the geometrical parameters reported. An example of this is the minimum required area; Aspen Plus® reports that value but Aspen HYSYS® omits this parameter. Overall, Aspen HYSYS® reports final results and very specific calculation, while Aspen Plus® provides values of other parameters that can lead to specific design adjustments. It is impossible to carry out a detailed geometric design of the heat exchanger from Aspen Plus® or Aspen HYSYS® process streams; these programs are limited to evaluate given geometric configurations. Instead, Aspen Exchanger Design and Rating®, as it was observed, has tools to do so.

The calculation is developed in Aspen Exchanger Design and Rating® considers several variables (design, evaluation, and simulation modes) that other modules lack; between these cost analysis, vibration analysis, and several heuristic rules are implemented. This program allows rigorous calculation as detail exchanger mechanical design, something very close to real equipment, largely because the software was designed to make detailed design evaluation of existing heat exchange equipment. Now a sensitivity analysis to observe the equipment performance can be done, considering that it is a distillation column condenser. Making a conceptual analysis can be considered that a change in top pressure can affect the energy consumption and therefore the amount of water required. Additionally, the scenario in which the n-propanol flow varies due to operational changes was studied.

Lesson 3　Chemical Engineering Design

Chemical engineering has consistently been one of the highest paidengineering professions. There is a demand for chemical engineers in many sectors of industry, including the traditional process industries: chemicals, polymers, fuels, foods, pharmaceuticals, and paper, as well as other sectors such as electronic materials and devices, consumer products, mining and metals extraction, biomedical implants, and power generation.

Starting from a vaguely defined problem statement such as a customer need or a set of experimental results, chemical engineers can develop an understanding of the important underlying physical science relevant to the problem and use this understanding to create a plan of action and set of detailed specifications, which, if implemented, will lead to a predicted financial outcome.

The creation of plans and specifications and the prediction of the financial outcome if the plans were implemented is the activity of chemical engineering design.

Design is a creative activity, and as such can be one of the most rewarding and satisfying activities undertaken byan engineer. The design does not exist at the start of the project. The designer begins with a specific objective or customer need in mind, and by developing and evaluating possible designs, arrives at the best way of achieving that objective; be it a better chair, a new bridge, or for the chemical engineer, a new chemical product or production process.

When considering possible ways of achieving the objective the designer will be constrained by many factors, which will narrow down the number of possible designs. There will rarely be just one possible solution to the problem, just one design. Several alternative ways of meeting the objective will normally be possible, even several best designs, depending on the nature of the constraints.

These constraints on the possible solutions to a problem in design arise in many ways. Some constraints will be fixed and invariable, such as those that arise from physical laws, government regulations, and engineering standards. Others will be less rigid, and can be relaxed by the designer as part of the general strategy for seeking the best design. The constraints that are outside the designer's influence can be termed the external constraints. These set the outer boundary of possible designs. Within this boundary there will be a number of plausible designs bounded by the other constraints, the internal constraints, over which the designer has some control; such as choice of process, choice of process conditions, materials, and equipment.

Economic considerations are obviously a major constraint on any engineering design: plants must make a profit.

Time will also be a constraint. The time available for completion of a design will usually limit the number of alternative designs that can be considered.

As the design develops, the designer will become aware of more possibilities and more constraints, and will be constantly seeking new data and evaluating possible design solutions.

The amount of work, and the way it is tackled, will depend on the degree of novelty in a design project. Development of new processes inevitably requires much more interaction with researchers and collection of data from laboratories and pilot plants.

Chemical engineering projects can be divided into three types, depending on the novelty involved:

(1) Modifications, and additions, to existing plant; usually carried out by the plant design group. Projects of this type represent about half of all the design activity in industry.

(2) New production capacity to meet growing sales demand, and the sale of established processes by contractors. Repetition of existing designs, with only minor design changes, including designs of vendor's or competitor's processes carried out to understand whether they have a compellingly better cost of production. Projects of this type account for about 45% of industrial design activity.

(3) New processes, developed from laboratory research, through pilot plant, to a commercial process. Even here, most of the unit operations and process equipment will use established designs. This type of project accounts for less than 5% of design activity in industry.

The majority of process designs are based on designs that previously existed. The design engineer very rarely sits down with a blank sheet of paper to create a new design from scratch, an activity sometimes referred to as "process synthesis". Even in industries such as pharmaceuticals, where research and new product development are critically important, the types of process used are often based on previous designs for similar products, so as to make use of well-understood equipment and smooth the process of obtaining regulatory approval for the new plant.

The design work required in the engineering of a chemical manufacturing process can be divided into two broad phases.

Phase 1: Process design, which covers the steps from the initial selection of the process to be used, through to the issuing of the process flowsheets; and includes the selection, specification, and chemical engineering design of equipment. In a typical organization, this phase is the responsibility of the process design group, and the work is mainly done by chemical engineers. The process design group may also be responsible for the preparation of the piping and instrumentation diagrams.

Phase 2: Plant design, including the detailed mechanical design of equipment, the structural, civil, and electrical design, and the specification and design of the ancillary services. These activities will be the responsibility of specialist design groups, having expertise in the whole range of engineering disciplines.

Other specialist groups will be responsible for cost estimation, and the purchase and procurement of equipment and materials.

Some of the larger chemical manufacturing companies have their own project design organizations andcarry out the whole project design and engineering, and possibly construction, within their own organization. More usually, the design and construction, and possibly assistance with start-up, are subcontracted to one of the international Engineering, Procurement and Construction (EPC) firms. The technical "know-how" for the process could come from the operating company or could be licensed from the contractor or a technology vendor. The operating company, technology provider, and contractor will work closely together throughout all stages of the project.

Words and Expressions

well-understood   *adj.* 良好理解, 充分了解

ancillary   ['ænsəˌleri]   *adj.* 辅助的, 补充的, 附属的

flowsheets   ['flouʃits]   *n.* 流程, 流程图, 工艺流程图

procurement   [prə'kjʊrmənt]   *n.* 采购

piping diagrams   *n.* 线路图, 管道图

know-how   *n.* 专门知识, 技能, 实际经验

instrumentation diagrams   *n.* 设备图

Notes

Engineering, Procurement and Construction (EPC):设计采购施工总承包,是工程总承包典型模式之一。

Exercises

1. Please list the classification of chemical engineering design depending on the novelty involved and describe each of them.

2. What should be considered in chemical engineering design? Why?

Reading Material: Basic Engineering Design

A basic engineering design report (BEDR) is often used at the end of the process design phase to collect and review information before beginning the plant design phase and detailed design of equipment, piping, plot layout, etc. The purpose of the BEDR is to ensure that all the information necessary for detailed design has been assembled, reviewed, and approved, so as to minimize errors and rework during detailed design. The BEDR also serves as a reference document for the detailed design groups and provides them with stream flows, temperatures, pressures, and physical property information. One of the most important functions of a basic engineering design report is to document the decisions and assumptions made during the design and the comments and suggestions made during design review meetings. These are often documented as separate sections of the report so that other engineers who later join the project can understand the reasons why the design evolved to its current form. A sample contents list for a basic engineering design report is given.

1. Process Description and Basis

1.1 Project Definition (customer, location, key feeds, and products)

1.2 Process Description (brief description of process flowsheet and chemistry, including block flow diagrams)

1.3 Basis and Scope of Design (plant capacity, project scope, design basis table)

2. Process Flow Diagrams

3. Mass and Energy Balances

3.1 Base Case Stream Data (stream temperature and pressure, mass flow and molar flow of each component in all streams, stream mass and molar composition, and total stream mass and molar flow, usually given as tables)

3.2 Modified Cases Stream Data (same data foreach variant design case, for example winter/summer cases, start of run/end of run, different product grades, etc.)

3.3 Base Case Physical Property Data (physical properties required by detailed design groups, such as stream density, viscosity, thermal conductivity, etc.)

4. Process Simulation (description of how the process was simulated and any differences between the simulation model and process flow diagram that detailed design groups need to understand)

5. Equipment List

6. Equipment Specifications

6.1 Pressure Vessels

6.2 Heaters

6.3 Heat Exchangers

6.3.1 Tubular

6.3.2 Air Cooled

6.4 Fluid Handling Equipment

6.4.1 Pumps

6.4.2 Compressors

6.5 Solid Handling Equipment

6.6 Drivers

6.6.1 Motors

6.6.2 Turbines

6.7 Unconventional or Proprietary Equipment

6.8 Instrumentation

6.9 Electrical Specifications

6.10 Piping

6.11 Miscellaneous

7. Materials of Construction (what materials are to be used in each section of the plant and why they were selected, often presented as a table or as a marked up version of the process flow diagram)

8. Preliminary Hydraulics (pump-and-line calculations of pressure drop used as a basis for sizing pumps and compressors)

9. Preliminary Operating Procedures (describe the procedures for plant start-up, shutdown, and emergency shutdown)

10. Preliminary Hazard Analysis (description of major materials and process hazards of the design)

11. Capital Cost Estimate (breakdown of capital cost, usually for each piece of equipment plus bulks and installation, usually given asa table or list)

12. Heat Integration and Utilities Estimate (overview of any pinch analysis or other energy optimization analysis, composite curves, table giving breakdown of utility consumption and costs)

13. Design Decisions and Assumptions (description of the most significant assumptions and selection decisions made by the designers, including references to calculation sheets for alterna-

tives that were evaluated and rejected)

14. Design Review Documentation

14. 1 Meeting Notes (notes taken during the design review meeting)

14. 2 Actions Taken to Resolve Design Review Issues (description of what was done to follow up on issues raised during the design review)

15. Appendices

15. 1 Calculation Sheets (calculations to support equipment selection and sizing, numbered and referenced elsewhere in the report)

15. 2 Project Correspondence (communications between the design team, marketing, vendors, external customers, regulatory agencies and any other parties whose input influenced the design)

Part 4　Safety, Environmental Protection and Engineering Ethics

Unit 1　Safety

After completing this unit, you should be able to:
1. Tell the difference between fire and explosion.
2. Tell the importance of safety for chemical process.
3. List the essential elements for combustion.

Lesson 1　Chemical Process Safety

It is reasonable to conclude that the growth of an industry is also dependent on technological advances. This is especially true in the chemical industry, which is entering an era of more complex processes: higher pressure, more reactive chemicals, and exotic chemistry. More complex processes require more complex safety technology. Many industrialists even believe that the development and application of safety technology is actually a constraint on the growth of the chemical industry.

As chemical process technology becomes more complex, chemical engineers will need a more detailed and fundamental understanding of safety. Since 1950, significant technological advances have been made in chemical process safety. Today, safety is equal in importance to production and has developed into a scientific discipline that includes many highly technical and complex theories and practices. Examples of the technology of safety include:

(1) Hydrodynamic models representing two-phase flow througha vessel relief;

(2) Dispersion models representing the spread of toxic vapor through a plant after a release;

(3) Mathematical techniques to determine the various ways that processes can fail and the probability of failure.

Recent advances in chemical plant safety emphasize the use of appropriate technological

114

tools to provide information for making safety decisions with respect to plant design and operation.

The word "safety" used to mean the older strategy of accident prevention through the use of hard hats, safety shoes, and a variety of rules and regulations. The main emphasis was on worker safety. Much more recently, "safety" has been replaced by "loss prevention. " This term includes hazard identification, technical evaluation, and the design of new engineering features to prevent loss. The subject of this text is loss prevention, but for convenience, the words "safety" and "loss prevention" will be used synonymously throughout.

Safety, hazard, and risk are frequently used terms in chemical process safety. Their definitions are

(1) Safety or loss prevention: the prevention of accidents through the use of appropriate technologies to identify the hazards of a chemical plant and eliminate them before an accident occurs.

(2) Hazard: a chemical or physical condition that has the potential to cause damage to people, property, or the environment.

(3) Risk: a measure of human injury, environmental damage, or economic loss in terms of both the incident likelihood and the magnitude of the loss or injury.

Chemical plants containa large variety of hazards. First, there are the usual mechanical hazards that cause worker injuries from tripping, falling, or moving equipment. Second, there are chemical hazards. These include fire and explosion hazards, reactivity hazards, and toxic hazards.

A successful safety program requires several ingredients, as shown in Figure 1. These ingredients are: ①System; ②Attitude; ③Fundamentals; ④Experience; ⑤Time; ⑥You.

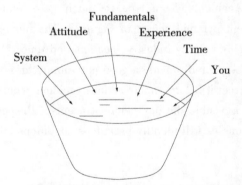

Figure 1　The ingredients of a successful safety program

First, the programneeds a system ① to record what needs to be done to have an outstanding safety program, ② to do what needs to be done, and ③ to record that the required tasks are done. Second, the participants must have a positive attitude. This includes the willingness to do some of the thankless work that is required for success. Third, the participants must understand

and use the fundamentals of chemical process safety in the design, construction, and operation of their plants. Fourth, everyone must learn from the experience of history or be doomed to repeat it. It is especially recommended that employees ① read and understand case histories of past accidents and ② ask people in their own and other organizations for their experience and advice. Fifth, everyone should recognize that safety takes time. This includes time to study, time to do the work, time to record results (for history), time to share experiences, and time to train or be trained. Sixth, everyone (you) should take the responsibility to contribute to the safety program. A safety program must have the commitment from all levels within the organization. Safety must be given importance equal to production. The most effective means of implementing a safety program is to make it everyone's responsibility in a chemical process plant. The older concept of identifying a few employees to be responsible for safety is inadequate by today's standards. All employees have the responsibility to be knowledgeable about safety and to practice safety. It is important to recognize the distinction between a good and an outstanding safety program.

· A good safety program identifies and eliminates existing safety hazards.

· An outstanding safety program has management systems that prevent the existence of safety hazards.

Agood safety program eliminates the existing hazards as they are identified, whereas an outstanding safety program prevents the existence of a hazard in the first place. The commonly used management systems directed toward eliminating the existence of hazards include safety reviews, safety audits, hazard identification techniques, checklists, and proper application of technical knowledge.

Property damage and loss of production must also be considered in loss prevention. These losses can be substantial. Accidents of this type are much more common than fatalities. This is demonstrated in the accident pyramid shown in Figure 2. The numbers provided are only approximate. The exact numbers vary by industry, location, and time. "No Damage" accidents are frequently called "near misses" and provide a good opportunity for companies to determine that a problem exists and to correct it before a more serious accident occurs. It is frequently said that "the cause of an accident is visible the day before it occurs. " Inspections, safety reviews, and careful evaluation of near misses will identify hazardous conditions that can be corrected before real accidents occur.

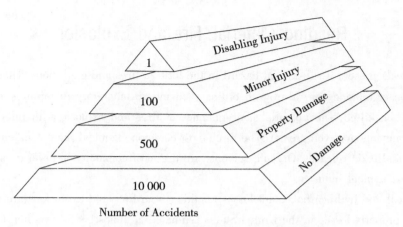

Figure 2   The accident pyramid

Words and Expressions

exotic　[ˌɡ'zɑːtˌk]　*adj.* 奇异的, 异国风味的

audits　['ɔːdˌts]　*n.* 审计, 稽核

throughout　[θruː'aʊt]　*adv.* 始终, 处处

release　[rˌ'liːs]　*v.* 释放, 放出, 放走

hydrodynamic　*adj.* 流体动力(学)的

pyramid　['pˌrəmˌd]　*n.* 金字塔, 锥体

doom　[duːm]　使……注定失败(或遭殃、死亡等)

synonymously　[sˌ'nɑːnˌməsli]　*adv.* 同义地

vessel relief　减压容器

hazardous　['hæzərdəs]　*adj.* 危险的, 有害的

Exercises

1. Match each word from the interview with the correct definition.

(1) compensation　　　　a. protect

(2) initial　　　　　　　b. small, specific change to something

(3) safeguard　　　　　c. probability, chance

(4) hazardous　　　　　d. existing at the beginning

(5) modificaiton　　　　e. payment for damage, harm, or loss

(6) likelihood　　　　　f. dangerous, risky

2. Write an account of an accident that you or someone you know had at school or work. Describe what happened, what caused the accident, and how, if at all, it could have been avoided.

Reading Material: Fire and Explosion

Chemicals present a substantial hazard in the form of fires and explosions. The combustion of one gallon of toluene can destroy an ordinary chemistry laboratory in minutes, and persons present may be killed. The potential consequences of fires and explosions in pilot plants and plant environments are even greater. The three most common chemical plant accidents are fires, explosions, and toxic releases. Organic solvents are the most common source of fires and explosions in the chemical industry.

Chemical and hydrocarbon plant losses resulting from fires and explosions are substantial, with yearly property losses in the United States estimated at almost $ 300 million (1997). Additional losses in life and business interruptions are also substantial. To prevent accidents resulting from fires and explosions, engineers must be familiar with

· The fire and explosion properties of materials,
· The nature of the fire and explosion process,
· Procedures to reduce fire and explosion hazards.

The essential elements for combustion are fuel, an oxidizer, and an ignition source. Fire, or burning, is the rapid exothermic oxidation of an ignited fuel. The fuel can be in solid, liquid, or vapor form, but vapor and liquid fuels are generally easier to ignite. The combustion always occurs in the vapor phase; liquids are volatized and solids are decomposed into vapor before combustion.

When fuel, oxidizer, and an ignition source are present at the necessary levels, burning will occur. This means a fire will not occur if ① fuel is not present or is not present in sufficient quantities, ② an oxidizer is not present or is not present in sufficient quantities, and ③ the ignition source is not energetic enough to initiate the fire.

Two common examples of the three components of the fire triangle are wood, air, and a match; and gasoline, air, and a spark. However, other, less obvious combinations of chemicals can lead to fires and explosions. Various fuels, oxidizers, and ignition sources common in the chemical industry are:

Fuels

Liquids: gasoline, acetone, ether, pentane;
Solids: plastics, wood dust, fibers, metal particles;
Gases: acetylene, propane, carbon monoxide, hydrogen.

Oxidizers

Gases: oxygen, fluorine, chlorine;
Liquids: hydrogen peroxide, nitric acid, perchloric acid;
Solids: metal peroxides, ammonium nitrite;

Ignition sources: Sparks, flames, static electricity, heat.

In the past the sole method for controlling fires and explosions was elimination of or reduction in ignition sources. Practical experience has shown that this is not robust enough-the ignition energies for most flammable materials are too low and ignition sources too plentiful. As a result, current practice is to prevent fires and explosions by continuing to eliminate ignition sources while focusing efforts strongly on preventing flammable mixtures.

The major distinction between fires and explosions is the rate of energy release. Fires release energy slowly, whereas explosions release energy rapidly, typically on the order of microseconds. Fires can also result from explosions, and explosions can result from fires. A good example of how the energy release rate affects the consequences of an accident is a standard automobile tire. The compressed air within the tire contains energy. If the energy is released slowly through the nozzle, the tire is harmlessly deflated. If the tire ruptures suddenly and all the energy within the compressed tire releases rapidly, the result is a dangerous explosion.

Some of the commonly used definitions related to fires and explosions are given in what follows. These definitions are discussed in greater detail in later sections.

Combustion or fire: Combustion or fire is a chemical reaction in which a substance combines with an oxidant and releases energy. Part of the energy released is used to sustain the reaction.

Ignition: Ignition of a flammable mixture may be caused by a flammable mixture coming in contact with a source of ignition with sufficient energy or the gas reaching a temperature high enough to cause the gas to autoignite.

Autoignition temperature (AIT): A fixed temperature above which adequate energy is available in the environment to provide an ignition source.

Flash point (FP): The flash point of a liquid is the lowest temperature at which it gives off enough vapor to form an ignitable mixture with air. At the flash point the vapor will burn but only briefly; inadequate vapor is produced to maintain combustion. The flash point generally increases with increasing pressure.

Fire point: The fire point is the lowest temperature at which a vapor above a liquid will continue to burn once ignited; the fire point temperature is higher than the flash point.

Flammability limits: Vapor-air mixtures will ignite and burn only over a well-specified range of compositions. The mixture will not burn when the composition is lower than the lower flammable limit (LFL); the mixture is too lean for combustion. The mixture is also not combustible when the composition is too rich, that is, when it is above the upper flammable limit (UFL). A mixture is flammable only when the composition is between the LFL and the UFL. Commonly used units are volume percent fuel (percentage of fuel plus air).

Explosion: An explosion is a rapid expansion of gases resultingin a rapidly moving pressure or shock wave. The expansion can be mechanical (by means of a sudden rupture of a pressurized vessel), or it can be the result of a rapid chemical reaction. Explosion damage is caused by

119

the pressure or shock wave.

Mechanical explosion: An explosion resulting from the sudden failure of a vessel containing high-pressure nonreactive gas.

Deflagration: An explosion in which the reaction front moves at a speed less than the speed of sound in the unreacted medium.

Detonation: An explosion in which the reaction front moves at a speed greater than the speed of sound in the unreacted medium.

Confined explosion: An explosion occurring within a vessel or a building. These are most common and usually result in injury to the building inhabitants and extensive damage.

Unconfined explosion: Unconfined explosions occur in the open. This type of explosion is usually the result of a flammable gas release. The gas is dispersed and mixed with air until it comes in contact with an ignition source. Unconfined explosions are rarer than confined explosions because the explosive material is frequently diluted below the LFL by wind dispersion. These explosions are destructive because large quantities of gas and large areas are frequently involved.

Boiling-liquid expanding-vapor explosion (BLEVE): A BLEVE occurs if a vessel that contains a liquid at a temperature above its atmospheric pressure boiling point ruptures. The subsequent BLEVE is the explosive vaporization of a large fraction of the vessel contents, possibly followed by combustion or explosion of the vaporized cloud if it is combustible. This type of explosion occurs when an external fire heats the contents of a tank of volatile material. As the tank contents heat, the vapor pressure of the liquid within the tank increases and the tank's structural integrity is reduced because of the heating. If the tank ruptures, the hot liquid volatilizes explosively.

Dust explosion: This explosion results from the rapid combustion of fine solid particles. Many solid materials (including common metals such as iron and aluminum) become flammable when reduced to a fine powder.

Shock wave: An abrupt pressure wave moving through a gas. A shock wave in open air is followed by a strong wind; the combination of shock wave and wind is called a blast wave. The pressure increase in the shock wave is so rapid that the process is mostly adiabatic.

Overpressure: The pressure on an object as a result of an impacting shock wave.

Words and Expressions

oxidizer ['ɑksˌdaɪzər] n. 氧化剂
Flammability limits n. 可燃度极限
ignition source [ɪgˈnɪʃn sɔːrs] n. 火源
rupture ['rʌptʃər] n. 破裂,断裂,爆裂
pentane ['pɛnˌteɪn] n. 戊烷,正戊烷
Deflagration [ˌdeflɑˈgreɪʃən] n. 爆燃(作用),焚烧

acetylene   [əˈsetəliːn]   *n.* 乙炔

Detonation   [ˌdet(ə)nˈeɪʃ(ə)n]   *n.* 爆炸, 起爆, 引爆

perchloric acid   高氯酸, 过氯酸

Dust explosion   粉尘爆炸

metal peroxides   [pəˈrɑkˌsaɪdz]   金属过氧化物

Boiling-liquid expanding-vapor explosion   沸液蒸气爆炸

ammonium nitrite   [əˈmoʊniəm ˈnaɪtraɪt]   *n.* 亚硝酸铵

Shock wave   (爆炸、地震等引起的) 冲击波

Autoignition   [ˌɔtɔˌgˈnɪʃən]   temperature 自燃点

adiabatic   [ˌædɪəˈbætɪk]   *adj.* 绝热的, 不传热的

Unit 2　Environmental Protection

After completing this unit, you should be able to:

1. Tell the importance of environmental protection for the chemical industry.
2. Tell the method to reduce the environmental effects of chemical industry.

Lesson 1　Chemical Engineering Environmental Protection

The business of the chemical industry is to produce and sell chemical products. Profitability is therefore an essential aim of production. However, no chemical process exists that produces only the product desired. Other substances not desired by the producer are also formed in the gas, liquid, or solid state. These are referred to as residues. Chemical production is thus a two-edged sword. On the one hand it manufactures products, i. e. , "goods" and on the other hand it produces residues, i. e. , "bads".

Another equally important aim is to reduce the environmental effects of these residues to a level that is acceptable. The chemical industry endeavors to operate the production process so as to minimize the entry of residues and waste products into the environment, both quantitatively and with respect to their hazard potential, and to utilize raw materials and energy with maximum efficiency. This growing awareness of the environment follows not only from the principle that the chemical industry is responsible for its actions, but also from the changed conditions governing the introduction of new production methods and the operation of existing plants. These include the following:

(1) The demanding international and national decrees and regulations covering environmental protection;

(2) In chemical companies, the correct allocation of costs required for "cleaning up" a given production unit to comply with regulations (e. g. , the purification of wastegas and wastewater and the disposal of waste materials from this unit);

(3) Difficulties in the disposal of waste materials due to the shortage of dumping space and the secondary costs of waste disposal plants;

(4) Costs of raw materials and energy;

(5) Increased public awareness of the importance of environmental protection;

(6) In addition, the processes and production plants of chemical technology must meet high standards of workplace protection and operational safety.

Thus, the chemical industry as a key industry can greatly contribute to the development of the concept of sustainable development.

It cannot offer a complete solution, but because of its knowledge and experience in handling substances, their processing and utilization as well as preparation and reuse it can mold important sections of this model. This is reflected in the following contributions:

(1) Improvement in value creation and productivity;

(2) Optimal management of the raw materials and energy;

(3) Environmentally friendly technologies as process improvements—as part of environmental protection and material cycles;

(4) Technical processes allowing, e. g. , safer disposal of waste.

A chemical production process generates residues in addition to the intended end product. These residues can be recycled. If recycling is impossible for technical or economic reasons, residues become waste. In addition, polluted waste water and waste gases can be formed. Disposal of wastes includes incineration and landfill disposal. Waste incineration generates secondary wastes in the form of slag and ashes, which have to be utilized or disposed of. Wastewater is produced in waste-gas scrubbing. The treatment of wastewater leads to the formation of sewage sludge and other sludges, which must then be disposed of as waste. Landfill leachate must also be treated, and once again this process can contribute to sludge formation. In this way, a complex residue/waste system results, and the solution of one problem results in the formation of secondary waste.

Environmental protection in the chemical industry is divided into product related and production related areas. Environmental protection related to products covers the development and production of environmentally friendly products (e. g. , paints, herbicide/pesticides, washing powder) and treatment of product wastes from processing and consumption. Environmental protection related to production covers the concept of the production-integrated environmental protection and additive environmental protection. The third concept covers production plant including the subsidiary areas of storing and packing of raw materials and products. It is necessary to construct and operate these, so that material loss does not lead to environmental damage. Furthermore this includes retention systems for contaminated water draining out of burning plant during fires and containment systems for cooling water to prevent contamination of groundwater or surface waters.

The environmental demands placed on the production process can be met with the aid of integrated or additive concepts.

The sustainable development model demands a new orientation of environmental politics. Therefore innovations in all areas of industrial environmental protection are the most effective tools. These will lead to an efficiency revolution. Its elements can be characterized as follows:

① Process innovation: Production of the same or similar products using less raw material and energy and giving lower pollutant output as waste gas, waste and polluted wastewater, i. e. , development of environmentally friendly processes.

② Material cycles: recycling of residues and product wastes can reduce use of resources.

The reduction and prevention of residues can be achieved by:

(1) Improving the chemical process with the aid of new synthesis routes. For example, in the production of aromatic amines, chemical reduction with iron chips by catalytic reduction by hydrogen. Shifting the equilibrium. The use of more favorable reaction conditions can cause the position of the equilibrium to be shifted so that one of the two components is almost 100% reacted. This can be achieved by using the second component in excess, by removing the product, or by using more favorable temperature or pressure.

(2) Improving selectively. A very effective method of reducing the amounts of residues and improving the yield is to increase the selectivity of the chemical reaction.

Examples of this include the following:

① Improvement of the selectivity of catalysts, e. g. , by using catalysts that lower the rate of an undesired side reaction;

② Maintenance of high catalytic activity, e. g. , by avoiding contact poisons or by developing simple reaction methods;

③ Optimization of reaction conditions, e. g. , by utilizing differences in the reaction kinetics of the main reaction and the side reaction, more favorable temperature profiles and residence times, or more suitable reactors;

④ Recycling of the side product (if the side reaction is reversible) developing simple reaction methods.

(3) Developing new catalysts, e. g. , in production of polypropylene without generation.

(4) Process optimization.

(5) Changing the reaction medium. If water is replaced by an organic solvent in syntheses, contamination of wastewater can often be drastically reduced. However, environmentally friendly solvent handling involves not only recovery of the solvent from liquid media but also prevention of losses to the atmosphere during storage, transport, production, and subsequent processing. This can be achieved, for example, by adsorptive recovery of solvents from the gas stream of wastewater using improved metal-organic catalysts.

(6) Using raw materials of higher purity.

(7) Replacing or eliminating auxiliaries that have a harmful effect on the environment (e. g. , chlorinated hydrocarbons).

Gaseous, liquid, or solid residues whose formation during the production process cannot be avoided, even under optimum operation conditions, can often be reused (Figure 1) by methods such as,

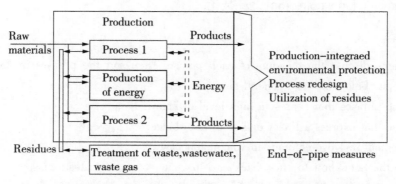

Figure 1   Approaches to environmental protection in chemical production

(1) Internal utilization, e. g. , of the auxiliaries employed in the process, by processing and recycling directly into the process. In the simplest case, the substance can be used directly after separation from the products or process streams (recovery of volatile components of solvents) . In other cases, a physical or chemical processing stage is necessary to remove impurities from the recycled components or convert them into a reusable form. Examples include:

① Recovery of sulfuric acid from waste sulfuric acid by concentration or decomposition and reprocessing;

② Recovery of organic solvents from solvent residues, solvent-product mixtures, aqueous solutions, and production residues;

③ Thermal decomposition of residues of chlorination processes to give pure hydrochloric acid.

(2) external utilization of the residue, i. e. , as a raw material for the manufacture of other products in a separate production plant.

Reutilization of residues by linking of production processes is not another form of residue disposal, but enables resources to be used as economically as possible. However, this does lead to additional interdependencies, which may decrease flexibility.

Words and Expressions

two-edged sword   双刃剑, 双刃的剑

herbicide   [ˈɜːrbɪˌsaɪd]   n. 除莠剂, 除草剂

Landfill   [ˈlændˌfɪl]   n. 垃圾填埋, 垃圾填埋地

pesticides   [ˈpɛstəˌsaɪdz]   n. 杀虫剂, 除害药物

leachate   [ˈliːtʃeɪt]   n. 沥滤液, 淋滤液

sludges   [ˈslʌdʒɪz]   n. 烂泥, 淤泥, 生活污物

polypropylene   [ˌpɑliˈproupəˌlin]   n. 聚丙烯

waste-gas scrubbing   废气洗涤

metal-organic catalysts   金属有机催化剂

incineration ［ˌɪnˈsɪnəˌreɪʃən］ *n.* 焚化，焚烧

Exercises

1. In pairs, discuss the meaning of each phrase. Then sort the phrases in Exercise band those in the box below into two general groups:

(1) those phrases that express a high level of knowledge;

(2) those that express a lesser degree of knowledge.

Arrangethem into two columns in your notebooks.

to have the know-how to know backwardsto know your classmatesto master

to know something like the back of your handto know the basics

2. Write sentences in your notebook about your own level of knowledge and expertise in specific areas of your field of study or work. Then compare and discuss your sentences with a classmate.

I am completely conversant with...

I know the basics of...

Reading Material: Production of Fatty Acid Methyl Esters

1. Fats and Oils: The Raw Materials of Oleochemistry

Fats and oils are triglycerides (i. e. , fatty acid esters of glycerol) . They are the starting materials for the production of fatty acid methyl esters, which are important intermediates in the production of fatty alcohols and surfactants by the oleochemical route, which has great ecological benefits. The fatty acid methyl esters are produced either by the esterification of fatty acids after hydrolysis of the triglycerides or by direct transesterification with methanol. The overall transesterification reaction is as follows:

$$
\begin{array}{c}
\underset{\text{Triglyceride}}{\left[\begin{array}{l} H_2C\!-\!O\!-\!\overset{\overset{O}{\|}}{C}\!-\!R_1 \\ HC\!-\!O\!-\!\overset{\overset{O}{\|}}{C}\!-\!R_2 \\ H_2C\!-\!O\!-\!\overset{\overset{O}{\|}}{C}\!-\!R_3 \end{array}\right]}
+\ \underset{\text{Methanol}}{3CH_3OH}
\ \underset{}{\overset{\text{base}}{\rightleftharpoons}}\
\underset{\text{Glycerol}}{\left[\begin{array}{l} H_2C\!-\!OH \\ HC\!-\!OH \\ H_2C\!-\!OH \end{array}\right]}
+\
\underset{\text{Methyl Esters}}{\left[\begin{array}{l} H_3C\!-\!O\!-\!\overset{\overset{O}{\|}}{C}\!-\!R_1 \\ H_3C\!-\!O\!-\!\overset{\overset{O}{\|}}{C}\!-\!R_2 \\ H_3C\!-\!O\!-\!\overset{\overset{O}{\|}}{C}\!-\!R_3 \end{array}\right]}
\end{array}
$$

This is actually the result of three consecutive reactions because the fatty acid groups of the triglyceride are rearranged sequentially. Transesterification can be carried out at a relatively low temperature (60 ℃) in the presence of an alkaline catalyst (e. g. , an alkali-metal alcoxide or hydroxide) .

The fat or oil raw material is not a pure triglyceride since significant amounts of other components are also present. Quantitatively, the most significant of these are free (unesterified) fatty acids (1% ~5%) , which neutralize the transesterification catalyst so that the consumption of catalyst can sometimes be very high. The catalyst dissolves in the joint product glycerol and must be removed as a salt at the end of the process.

Minimization of catalyst consumption is desirable for environmental reasons: therefore, the free fatty acids contained in the oil must be neutralized.

2. Low-pressure Transesterification—Old Batch Process

In the old batch transesterification process, neutralized oil is used. The oil is treated with aqueous sodium hydroxide to neutralize the free fatty acids (Figure 1) . The aqueous phase, which contains a high proportion of the soaps formed (soap stock) , is then removed by means of separators, and the remaining soap residues are removed from the oil by washing with water several times (usually twice) . These soap residues are then separated; the wash water and the aqueous phase from the first stage, which has a high soap content, are bulked together and treated with sulfuric acid. The soaps are thereby split to form fatty acids and sodium sulfate. The fatty acids can be separated as an organic phase by decantation. The water, which contains sodium sulfate and organic impurities, must be treated in a wastewater treatment plant. The fatty acids produced are a low-value joint product.

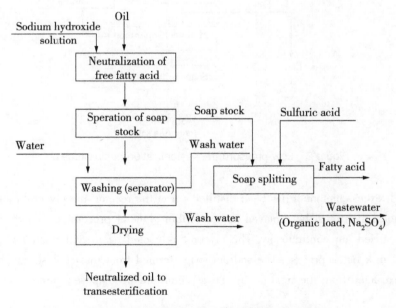

Figure 1　Flow diagram of alkaline refining

The oil, which is free of fatty acids, is then dried and transesterified (Figure 2) . The batch transesterification process represents the usual state of technology. The process is operated in as simple a way as possible. Sodium hydroxide dissolved in methanol is used as the catalyst for methanolysis. A high proportion of the sodium hydroxide is converted to sodium methoxide in the solution, liberating water:

$$NaOH + CH_3OH \rightleftharpoons NaOCH_3 + H_2O$$

The dried oil is mixed with methanol and the catalyst solution, and the reaction mixture is pumped into a stirred tank. The reaction is completed there at 80 ℃ as a batch operation. The methanol that evaporates is completely condensed and returned to the reactor.

At the end of the reaction, excess methanol is partly distilled off. The mixture of fatty acid methyl ester and glycerol is pumped into a settling tank to allow phaseseparation. After standing for several hours, the lower phase (glycerol phase) is separated. The upper ester phase is removed and temporarily stored before further processing (e. g. , hydrogenation to fatty alcohols) .

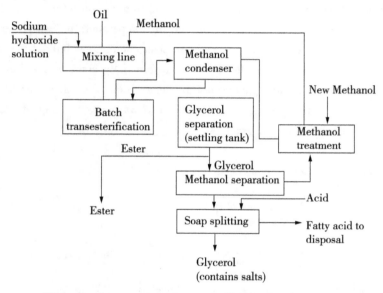

Figure 2   Low-pressure transesterification-batch process

Although excess methanol has been distilled out of the reactor, the glycerol phase still contains methanol, which must be removed before further glycerol processing. Removal of residual methanol is carried out continuously. The glycerol is cooled again and treated with acid in a stirred vessel in a batch process. The sodium soaps formed from the catalyst, which are almost completely extracted from the mixture by the glycerol, are decomposed, forming fatty acid and sodium salt.

The fattyacids can be separated after a short time as the upper phase.

The old batch transesterification process has the following disadvantage:

(1) Transesterification is carried out in only one step, so that a large excess of methanol is necessary to give a highconversion efficiency.

(2) The catalyst used is sodium hydroxide dissolved in methanol. Water is formed in the course of alcoxide formation. In the presence of water, the sodium hydroxide present reacts very quickly with fatty acid methyl esters to form sodium soaps. Under these conditions, the soaps formed are not effective catalysts.

For this reason, large quantities of catalyst must be used. This leads to product loss and to large quantities of fatty acids in the crude glycerol after soap splitting. Furthermore, the crude glycerol then has a high salt content. This amount of salt must be removed in subsequent processing and disposed of as waste.

3. New Method of Low-pressure Transesterification

The modified process has the following aims:

(1) Continuous operation;

(2) Utilization of the free fatty acids contained in the oil;

(3) Prevention or minimization of waste products and residues;

(4) Optimum utilization of the raw materials used, especially reduction in the amount of catalyst and the amount of excess methanol required;

(5) Heat recovery and optimization of energy consumption.

Neutralization of Oils by Pre-Esterifcation. The free fatty acids present in the raw material are esterified with methanol. The proton-catalyzed esterification can be carried out homogeneously. However, the result is that some of the acidic catalyst is also esterified, and the ester and catalyst must then be separated by a water wash. This results in wastewater that sometimes contains toxic substances. To avoid this problem, a heterogeneous catalyst is used (see Figure 3) consisting of a strongly acid macroporous ionexchange resin arranged in a fixed-bed reactor. The reaction temperature is between 60 ℃ and 80 ℃.

In pre-esterification, not only are the fatty acids in the crude oil esterified, but so are the fatty acids produced on splitting the soaps formed in the transesterification process. Recycling leads to an increased yield of methyl ester. After pre-esterification, excess methanol is separated and removed together with the reaction water.

The neutralized oil (mixture of triglycerides and methyl esters) is transesterified after addition of methanol and catalyst. Transesterification leads to the formation of glycerol and fatty acid methyl esters. A marked solubility gap exists in the three-component system glycerol-methyl ester-methanol, so that a glycerol phase and an ester phase are formed. This two-phase system gives the possibility of a multistage process. By removing the product glycerol (GLY, see Eq. 1), a high yield of methyl ester (ME) with a reduced consumption of methanol (MeOH) is achieved

$$K = \frac{[ME]^3 \times [GLY]}{[MEOH]^3 \times [TRI]} \tag{1}$$

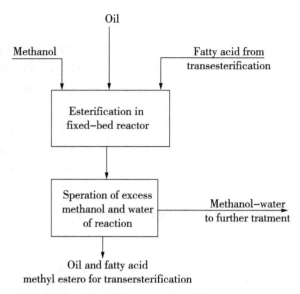

Figure 3　Flow diagram of pre-esterification

Transesterification is carried out in two stages, the glycerol phase formed being removed after each stage.

The reaction is carried out in tubular reactors with a flow rate that ensures good mixing of the two-phase reaction mixture at the beginning and end of the reaction. Since the amount of back mixing is low in these tubular reactors, short reaction times with low catalyst concentrations can be achieved at ca. 80 ~ 90 ℃.

With a new catalyst (alkali-metal oxide), smaller catalyst concentrations can be used than is the case with sodium hydroxide. With suitable reactor design and choice of catalyst, the salt content of the glycerol produced is decreased by 50%, and smaller quantities of fatty acids are formed in the splitting reaction. Immediately after transesterification, excess methanol contained in both liquid phases is removed with the aid of a multistage falling-film evaporator. The multistage system enables heat from the hot methyl ester and glycerol to be utilized in the initial evaporation stages.

The recovered methanol at atmospheric pressure is fed to the methanol distillation stage as a vapor.

The soaps in the glycerol from which the methanol has been removed are continuously split with acid. Consumption of acid is minimized by pH control. Concentrated acid is used to minimize the dilution of glycerol with water.

The ester phase, from which the methanol has been removed, still contains undissolved, highly dispersed residual glycerol, which is removed after cooling with the aid of coalescence separators. The use of such equipment avoids the need to wash the esters with water. A highly concentrated crude glycerol is thus obtained, which leads to reduced energy consumption in glycerol processing.

4. Ecological Aspects of the New Low-pressure Transesterification Process with Pre-esterification

In the earlier process, refined oil was used. The waste streams produced in the oil refining process are completely avoided by use of preesterification. Moreover, the waste fatty acid produced in the low-pressure transesterification process can be reutilized.

The reaction is catalyzed by a macroporous cation exchanger. This catalyst has a very long service life and can be disposed of without problems.

By carrying out the transesterification reaction in several stages, the process can be carried out with a lesser excess of methanol. This leads to an energy saving in the methanol recovery and methanol purification processes. The use of a continuous process enables heat to be recovered during the recovery of methanol from the ester and glycerol.

Impure methanol is fed into the methanol purification stage as a vapor, so that the energy consumption in methanol distillation is very low.

The use of an alkali-metal alcoxide in combination with the new reactor design has the advantage of lower catalyst consumption. Since the catalyst is separated along with the by-product glycerol, this means that the amounts of salt produced in a subsequent glycerol processing stage are greatly reduced.

During the process, crude glycerol is diluted as little as possible (use of coal escence separators) to give reduced energy consumption in glycerol processing and to minimize the amount of wastewater produced.

Words and Expressions

Oleochemistry　　*n.* 油脂化学

unesterified　　*adj.* 未酯化的

triglycerides　[traɪˈglɪsəˌraɪdz]　*n.* 甘油三酯

decantation　[ˌdiˌkænˈteɪʃən]　*n.* 缓倾(法), 倾析(法)

transesterification　[trænsəstərəfɪˌkeɪʃən]　*n.* 酯交换反应

methanolysis　[meθəˈnɒlˌsɪs]　*n.* 甲醇分解(作用)

alcoxide　　*n.* 烃氧基金属

methyl ester　[ˈmeθəl ˈestər]　甲酯, 脂肪酸甲酯

131

Unit 3　Engineering Ethics

After completing this unit, you should be able to:

1. List the things regard the importance of trust worthiness in engineers.
2. List the four catagories of the task of ethical analysis.

Lesson 1　Engineering Ethics

A person's profession is a part of her personal identity. According to several prominent accounts, engineering is a profession, although the absence in a jurisdiction of a requirement for registration in order to practice engineering weakens its professional status in that jurisdiction. Engineering codes and other statements from leaders of the engineering profession impose on engineers an obligation to promote the public good, sometimes interpreted as well-being and also as welfare or quality of life. Promoting the well-being of the public includes not engaging in professionally prohibited actions, preventing harm to the public, and actively promoting the public s well-being. In designing for well-being, engineers must keep in mind the social context of engineering and technology, and the need for a critical attitude toward technology.

The first task of ethical analysis is to sort out the issues in a case into four categories: factual issues, conceptual issues, application issues, and moral issues. The line-drawing method is a way of comparing a controversial situation with uncontroversial (paradigm) ones in order to determine what should be said about the controversial situation. The creative-middle-way approach is a way of resolving moral problems involving competing moral demands by coming up with courses of action that satisfy as many moral demands as possible. What are sometimes called moral theories or approaches to moral thinking are attempts to identify the fundamental idea(s) in common morality. They are not always necessary for resolving a moral problem, but, when they are, it is better to use more than one approach. The utilitarian approach finds the fundamental idea of common morality to be the imperative to maximize overall well-being. There are several ways of applying the utilitarian approach. The respect for persons (RP) approach finds the basic idea of common morality to be the imperative to act so as to respect humans as free and equal moral agents. There are several ways of applying the RP approach. The virtue ethics approach finds the basic idea of common morality to be the imperative to act in the way the virtuous person would act. It supplies concepts for understanding moral motivation and development and gives guidance when moral and professional rules provide insufficient direction.

Responsibility has to do with accountability, both for what one does in thepresent and future and for what one has done in the past. The responsibilities of engineers require not only adhering to regulatory norms and standard practices of engineering but also satisfying the standard of reasonable care. Engineers can expect to be held accountable, if not legally liable, for intentionally, negligently, and recklessly caused harms. Responsible engineering practice requires good judgment, not simply following algorithms. A good test of engineering responsibility is the question: What does an engineer do when no one is looking? This makes evident the importance of trust in the work of engineers. Responsible engineering requires taking into account various challenges to appropriate action, such as blind spots, normalized deviance, bounded ethicality, uncritical acceptance of authority, and groupthink.

Communication and culture are vital components within the organization, and employees need to understand them well. Employees should take advantage of organizational resources in order to enhance their own integrity and independence. Organizational and management practices may be unchanged for years, which can result in blind spots, or obstacles to ethical decision-making. Understanding the obstacles and remedies for these obstacles can improve the organization' s communication and ethical decision-making. Many organizations hire an ethics and compliance officer to study inappropriate policies and procedures and to assist employees in appropriate communication and daily ethical choices at work. Engineers and managers have different perspectives, both legitimate, and it is useful to distinguish between decisions that should be made by managers, or from a management perspective, and decisions that should be made by engineers, or from an engineering perspective. Differences of opinion can be expected within the organization between engineers themselves and between engineers and management. Careful verbal and written communication can be utilized to work through disagreements. Whistle blowing sometimes becomes a necessary option for an employee when other avenues of communication fail. An employee should explore numerous ways of solving an organizational problem before whistleblowing. However, new federal regulations are in place to assist employees who believe they have exhausted all other means of solving the workplace problem.

There are issues which regards importance of trustworthiness in engineers: honesty, confidentiality, intellectual property, expert witnessing, public communication, and conflicts of interest. Forms of dishonesty include lying, deliberate deception, withholding information, and failure to seek out the truth. Dishonesty in engineering research and testing includes plagiarism and the falsification and fabrication of data. Engineers are expected to respect professional confidentiality in their work. Integrity in expert testimony requires not only truthfulness but also adequate background and preparation in the areas requiring expertise. Conflicts of interest are especially problematic because they threaten to compromise professional judgment.

Engineers impose risks on the public in design and in management of engineered systems andinfrastructure and have an obligation to assess and manage these risks. Engineers and risk experts define risk as the product of the probability of a harm and the magnitude of that harm. In quantifying risks, engineers and risk experts have traditionally considered only harms that are relatively easily quantified, such as economic losses, bodily injury, or the number of lives lost. In a new version of the way engineers and risk experts deal with risk, the capabilities approach focuses on the broader effects of risks and disasters on the capabilities of people to live the kinds of lives they value. The public is concerned about informed consent and the just distribution of risk. Engineers have techniques for estimating the causes and likelihood of harm, but their effectiveness is limited.

The modern environmental movement began in the 1960s and 1970s. It has not only influenced law and policy throughout the world but also engineering practice. Life cycle analysis (or life cycle assessment) (LCA) is a useful basis for many types of engineering that affect the environment. Many formulations of the ideal of sustainability mandate care for the earth's resources for the sake of present and future generations and a economic development in less economically developed societies. Business firms exhibit a variety of attitudes toward environmental concerns, but some give evidence of a genuinely progressive attitude. The professional virtue of respect for the natural world should be cultivated, because it can motivate environmental concern on the part of engineers. Student members of organizations such as Engineers for a Sustainable World have probably already developed this virtue.

Some progress has been made in establishing international technical standards. Whether engineers worldwide think (or should think) of themselves as professionals is more controversial. Economic, cultural, and social differences between countries sometimes produce boundary-crossing problems for engineers. Solutions to these problems must avoid absolutism and relativism and should find a way between moral rigorism and moral laxism. Applying the standards of one s own country without modification or uncritically adopting the standards of the host country in which one is working are rarely satisfactory solutions to the moral issues that arise in international engineering. Solutions involving creative middle ways are often particularly useful. Engineering work in the international arena can raise many ethical issues, including exploitation, bribery, extortion, grease payments, nepotism, excessive gifts, paternalism, and paying taxes in a country where taxes are negotiable.

Words and Expressions

line-drawing   *n.* 线条画, 白描, 划界法

laxism   [ˈlakˌsizəm]   *n.* 宽纵说, 散漫主义

controversial   [ˌkɑntrəˈvɜrʃ(ə)l]   *adj.* 有争议的

exploitation   [ˌeksplɔɪˈteɪʃ(ə)n]   *n.* 开发, 开采, 利用

creative-middle-way   *n.* 创造性的中间方法

bribery   [ˈbraɪbəri]   *n.* 贿赂, 受贿, 行贿

utilitarian   [juˌtɪlɪˈteriən]   *n.* 功利主义者

extortion   [ɪkˈstɔrʃ(ə)n]   *n.* 勒索, 强夺

groupthink   [ˈgruːpθɪŋk]   *n.* 小集团思想

grease payments　润滑费

ethicality　［eθɪˈkælɪtɪ］　*n.* 伦理性, 道理性

nepotism　［ˈnepəɪtɪzəm］　*n.* 裙带关系, 任人唯亲

whistleblowing　［wɪslbˈləʊɪŋ］　*na.* 告密, 揭发

paternalism　［pəˈtɜːrnəlɪzəm］　*n.* 家长作风, 专制

rigorism　［ˈrɪɡərɪzəm］　*n.* 严格主义

Life cycle analysis　［laɪf ˈsaɪkl əˈnæləsɪs］　*n.* 生命周期分析法

Notes

Life cycle analysis:生命周期分析法是运用生命周期分析矩阵,根据企业的实力和产业的发展阶段来分析评价战略的适宜性的一种方法。利用它有助于战略选择,可以缩小选择的范围,做到有的放矢。生命周期矩阵的横坐标代表产业发展的阶段——幼稚、成长、成熟、衰退,纵坐标代表企业的实力,分为五类——主导、较强、有利、维持、脆弱。

Engineers for a Sustainable World:可持续世界工程师(ESW)是一个工程项目团队,致力于为本地和全球可持续性挑战打造创新、持久的解决方案。自 2002 年在康奈尔大学成立以来,ESW 已发展成为一个国际非营利网络,拥有 50 多个学院、大学和城市分会,4 000多名学生、教师和专业人员。ESW 康奈尔大学由三个子团队组成:生物燃料、可再生能源设计(RED)和太阳能解决方案(SPS)。

Exercises

Read thetext and match each underlined verb with the most appropriate less formal synonym. Use the infinitive form of the verb.

Computers <u>assist</u> mechanical engineers by <u>performing</u> accurate and efficient computation, and by <u>permitting</u> the modeling and simulation of a new design as well as <u>facilitating</u> changes to an existing design. Computern-Aided Design (CAD) and Computer-Aided Manufacturing (CAM are <u>employed</u> for design data processing and for <u>converting</u> a design into a product.

(1) allow　　　　　　　　(4) help

(2) carry out　　　　　　　(5) make easier

(3) change　　　　　　　　(6) use

Reading Material: Sustainability

Scientists, engineers, and the government are publicly expressing urgent concern about the need to address the challenges of sustainable scientific and technological development. Global warming, for example, raises concern about glacial meltdown and consequent rising ocean levels threatening coastal cities. A related concern is the lowering of levels of freshwater in the American West as a result of lowered levels of accumulated mountain snow. In Joe Gertner's "The Future Is Drying Up", Nobel laureate Steven Chu, director of the Lawrence Berkeley National

Laboratory, is cited as saying that even optimistic projections for the second half of the twenty-first century indicate a 30% to 70% drop in the snowpack level of the Sierra Nevada, provider of most of northern California's water. Gertner goes on to discuss other likely freshwater problems that will have to be faced by Western states as a result of both global warming and the consumption needs and demands of an increasing population. He also outlines some of the efforts of engineers to address these problems aggressively now rather than wait until it is too late to prevent disaster.

Most engineering society codes of ethics do not make direct statements about the environmental responsibilities of engineers. However, in 2007, the NSPE joined the ranks of engineering societies that do. Professional Obligations, provision 2 reads, "Engineers shall at all times strive to serve the public interest". Under this heading, there is a new entry: "Engineers are encouraged to adhere to the principles of sustainable development in order to protect the environment for future generations". Footnote 1 addresses the conceptual question of what is meant by "sustainable development":

"Sustainable development" is the challenge of meeting human needs for natural resources, industrial products, energy, food, transportation, shelter, and effective waste management while conserving and protecting environmental quality and the natural resource base essential for future development.

Although this definition of sustainable development leaves many fundamental conceptual and value questions in need of further analysis (e. g. , What are human needs? What is meant by"environmental quality"?) , it provides a general framework for inquiry. It also identifies a variety of fundamental areas of concern (e. g. , food, transportation, and waste management). Of course, responsibilities in these areas do not fall only on engineers. Government officials, economists, business leaders, and the general citizenry need to be involved as well. Thus, a basic question relates to how those who need to work together might best do so and what role engineers might play. We offer three illustrations for discussion. The first is an early effort to involve students from different disciplines in a project that supports sustainable development. The second is the recent proliferation of centers and institutes for sustainability on college campuses throughout the country. The third is service learning opportunities in support of sustainable design and development.

1. Renewable Energy

Dwayne Breger, a civil and environmental engineer at Lafayette College, invited junior and senior engineering, biology, and environmental science students to apply to be on an interdisciplinary team to design a project that would make use of farmland owned by Lafayette College in a way that supports the college mission. Twelve students were selected for the project: two each from civil and environmental engineering, mechanical engineering, chemical engineering, and bachelor of arts in engineering, plus three biology majors and one in geology and environmental geosciences. These students had minors in areas such as economics and business, environmental science, chemistry, government, and law. The result of the project was a promising design for a biomass farm that could provide an alternative, renewable resource for the campus steam plant.

Professor Breger regards projects such as this as providing important opportunities for students to involve themselves in work that contributes to restructuring our energy use toward sustainable resources. ABET' s Engineering Criteria 2000 for evaluating engineering programs includes the requirement that engineering programs demonstrate that their graduates have "an understanding of professional and ethical responsibility", "the broad education necessary to understand the impact of engineering solutions in a global and societal context", and "a knowledge of contemporary issues". Criterion 4 requires that students have "a major design experience" that includes consideration of the impact on design of factors such as economics, sustainability, manufacturability, ethics, health, safety, and social and political issues. Discuss how the Lafayette College project might satisfy criterion 4, especially the ethical considerations.

2. Academic Centers for Sustainability

Historically, joint research in colleges and universities is done within separate disciplines rather than in collaboration with other disciplines. Thus, biologists collaborate with other biologists, chemists with other chemists, economists with other economists, and political scientists with other political scientists. The recent emergence of centers and institutes for sustainability represents a significant and important break from that tradition.

In September 2007, the Rochester Institute of Technology initiated the Golisano Institute for Sustainability. Noting that it is customary for new programs tobe run by just one discipline, Nabil Nasr, the institute director, comments. But the problem of sustainability cuts across economics, social elements, engineering, everything. It simply cannot be solved by one discipline, or even by coupling two disciplines.

Dow Chemical has recently given the University of California at Berkeley $10 million to establish a sustainability center. Dow' s Neil Hawkins says, "Berkeley has one of the strongest chemical engineering schools in the world, but it will be the MBA s who understand areas like microfinance solutions to drinking water problems". The center is in Berkeley' s Center for Responsible Business, directed by Kellie A. McElhaney. Commercialization of research undertaken by students and professors is expected. However, McElhaney notes, "Commercialization takes forever if the chemical engineers and the business types do not coordinate. So think how much easier it will be for chemistry graduates to work inside a company if they already know how to interact with the business side".

Discuss how considerations of ethics might enter into the collaborative efforts of centers and institutes for sustainability.

3. Service Learning Opportunities

The first two issues of the recently launched International Journal for Service Learning feature three articles promoting the notion that service learning projects can provide hands-on opportunities to undertake sustainable design and development. In "Service Learning in Engineering and Science for Sustainable Development", Clarion University of Pennsylvania physicist Joshua M. Pearce argues that undergraduates should have opportunities to become involved in projects that apply appropriate technologies for sustainable development. Especially concerned with alleviating poverty in the developing world, Pearce argues, The need for development is as great as it has ever been, but future development cannot simply follow past models of economic activity, which tended to waste resources and produce prodigious pollution. The entire world is now paying to clean up the mess and enormous quantities of valuable resources have been lost for future generations because of the Western model of development. For the future, the entire world population needs ways to achieve economic, social, and environmental objectives simultaneously.

He cites successful projects in Haiti and Guatemala that make use of readily available materials in the localesin which they have been undertaken.

In"Learning Sustainable Design through Service", Stanford University PhD students Karim Al-Khafaji and Margaret Catherine Morse present a service learning model based on the Stanford chapter of Engineers for a Sustainable World to teach sustainable design. They illustrate this model in discussing a Stanford project in the Andaman Islands that focused on rebuilding after the December 26, 2004, earthquake and tsunami.

Design for a Sustainable World, that seeks to develop students iterative design skills, project management and partnership-building abilities, sustainability awareness, cultural sensitivity, empathy, and desire to use technical skills to promote peace and human development.

Help developing communities ensure individuals human rights via sustainable, culturally appropriate, technology-based solutions.

Increase Stanford University s stewardship of global sustainability.

In "Sustainable Building Materials in French Polynesia", John Erik Anderson, Helena Meryman, and Kimberly Porsche, graduate students at the University of California at Berkeley' s Department of Civil and Environmental Engineering, provide a detailed, technical description of a service learning project designed to assist French Polynesians in developing a system for the local manufacturing of sustainable building materials.

Words and Expressions

laureate　[ˈlɔriət]　n. 荣誉获得者, 获奖者
geosciences　[ˌdʒiːəʊˈsaɪəns]　n. 地球科学
localesin　局部的
ABET　美国工程与技术认证委员会
snowpack　[ˈsnoʊˌpæk]　n. 积雪场, 积雪层
Criterion　[kraɪˈtɪriən]　n. 标准, 准则, 原则
provision　[prəˈvɪʒn]　n. 规定, 条款

Dow Chemical　［daʊ ˈkemɪkl］　美国陶氏化学公司

Footnote　［ˈfʊtnoʊt］　*n.* 脚注,注脚

microfinance　［ˈmaɪkroʊfaɪˌnæns］　*n.* 小额信贷,微观金融

citizenry　［ˈsɪtɪzənri］　*n.* 全体市民(或公民)

tsunami　［tsuːˈnɑːmi］　*n.* 海啸,海震

Notes

1. ABET(Accrediation Board for Engineering and Technology):美国工程与技术认证委员会,创建于 1932 年,是专门从事工程、技术、电脑、应用科学的学术机构工程及技术教育认证的,具有公正性和权威性。

2. Dow Chemical:美国陶氏化学公司,创建于 1897 年,是一家以科技为主的跨国性公司,位居世界化学工业界第二名的国际跨国化工公司(美国杜邦公司居第一位)。

Part 5　Surface Chemistry and Surfactants

Unit 1　Surface Chemistry

After completing this unit, you should be able to:
1. List the associated structures of surfactants.
2. List the solution properties of surfactants.
3. Tell the reason of adsorption and wetting phenomenon of surfactant.

Lesson 1　Solution Properties of Surfactants

Because of their chemical composition, surfactants exhibit a "love-hate" relationship with most solvents, particularly water, which results in a tug-of-war between forces tending toward a comfortable accommodation within a given solvent environment and a driving "desire" to escape to a more energetically favorable situation. Anthropomorphically, surfactants seem to feel that the grass is always greener on the other side of the fence and, as a result, spend much of their time sitting on the "fence" between phases. That fence-sitting characteristic is manifested in many ways including adsorption at various phase interfaces, interaction with materials dissolved or dispersed in the liquid phase, and the formation of so so-called association colloids.

The adsorption of surfactants at interfaces will be discussed in later lesson. The formation of the various associated structures such as micelles, vesicles, membranes, and liquid crystals may be seen as a fundamental characteristic of a given molecular structure in a defined environment. The exact solution behavior of a surfactant will depend on a number of internal (molecular) and external factors, which will be discussed in turn. At this point it will be useful to take a look at the possibilities open to surface active molecules in the various environments that may be encountered.

The simplest class of association colloids is the micelle. The number of publications related to micellization, micelle structures, and the thermodynamics of micelle formation is enormous. Extensive interest in the self-association phenomenon is evident in such wide-ranging chemical and technological areas as organic and physical chemistry, biochemistry, polymer chemistry, pharmaceuticals, petroleum and minerals processing, cosmetics, and food science. Even with the vast amount of experimental arid theoretical work devoted to the understanding of the aggregation of surface active molecules, no complete theory of model has emerged that unambiguously satisfies all of the evidence and the interpretations of the evidence for micelle formation.

The solution behavior of surfactants reflects the unique "split personality" of such species. The thermodynamic pushing and pulling that occurs in aqueous solution (or nonaqueous solution, for that matter) result from a complex combination of effect including ① interactions of the hydrocarbon portions of the molecules with water, ② attractive interaction among hydrocarbon tails on separate molecules, ③ interactions among solvated head groups (generally, repulsive) and between the head groups and co-ions, in the case of ionic materials, and ④ geometric and packing constraints deriving from the particular molecular structure involved.

It is generally accepted that most surface active molecules in aqueous solution can aggregate to form micellar structures with an average of from 30 to 200 molecules in such a way that the hydrophobic portions of the molecules are associated and mutually protected from extensive contact with the bulk of the water phase. Not so universally accepted are some of the ideas concerning micellar shapes, the nature of the micellar interior, surface "roughness", the sites of adsorption (or solubilization) into (or onto) micelles, and the size distribution of micelles in a given system. Although ever more sophisticated experimental techniques continue to provide new insights into those questions, we still have a lot to learn.

Early in the study of surfactants, it became obvious that the solution properties of such materials were unusual and could change dramatically over very small concentration ranges. The measurement of solution properties such as surface tension, electrical conductivity, or light scattering as a function of surfactant concentration will produce curves that exhibit relatively sharp discontinuities to a comparatively low concentration (Figure 1). The sudden change in a measured solution properly is interpreted as indicating a change in the nature of the solute species affecting the measured quantity. In the measurement of conductivity (curve A), the break is associated with an increase in the mass per unit charge of the conducting species. For light scattering (curve B), the change in solution turbidity indicates the appearance of a scattering species of significantly greater size than the monomeric solute. These and many other types of measurement serve as evidence for the formation of aggregates or micelles in solutions of surfactants at relatively well-defined concentrations.

Results of early studies of such solution properties were classically interpreted in terms of a spherical association of surfactant molecules—the micelle. The structure was assumed to be an aggregate of from 50 to 100 molecules with a radius approximately equal to the length of the hydrocarbon chain of the surfactant. The interior of the micelle was described its being essentially hydrocarbon in nature, while the surface consisted of a shell of the head groups and associated counter ions, solvent molecules, etc.

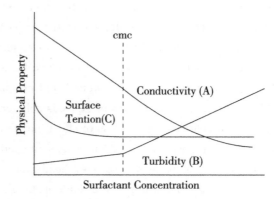

Figure 1　Typical varitations in solution physical properties with surfactant concentration

Modem studies using techniques unavailable just a few years ago have produced more detailed information about the submicroscopic nature of micelles. For example, they are not static species; they may be very dynamic with a constant, rapid interchange of molecules between the aggregates and the solution phase. It is therefore unreasonable to assume that surfactant molecules pack into a micelle in such an orderly manner as to produce a smooth, perfectly uniform surface structure. If one could photograph a micelle with ultra-high speed film, freezing the motion of the molecules, the picture would certainly show an irregular molecular cluster more closely resembling a cocklebur than a billiard ball.

Although the classical picture of a micelle is that of a sphere, most evidence suggests that spherical micelles are probably the exception rather than the rule. Due to geometric packing requirements (to be discussed below) ellipsoidal, disk-shaped, and rod like structures may be the more common micellar shapes. However, for the purpose of providing a basic concept of micelles and micelle information for the nonspecialist, the spherical model remains a useful and meaningful tool.

Leaving aside for the moment questions concerning particular concepts and models of micelles, it is known that the processe of micelle formation is thermodynamically driven. It is important, therefore, to have a general understanding of the rules that govern the process.

Words and Expressions

monomeric　[ˌmɒnə'merɪk]　*adj.* 单体的, 单节显性的

vesicle　['vesɪkl]　*n.* 囊, 泡

surfactant　[sɜː'fæktənt]　*n.* 表面活性剂(的)

roughness　[rʌfnəs]　*n.* 粗糙, 粗糙度

turbidity　[tɜː'bɪdəti]　*n.* 浊度, 不透明, 混乱

tug-of-war　激烈的竞争

micelle　[ma｜'sel]　*n.* 胶束, 胶粒, 胶囊

self-association　自缔合(作用)

micellization　胶束形成, 胶束化(作用)

well-define　轮廓清楚的, 定义, 明确

Exercises

1. Please judge whether the following statement is correct.

(1) Cloud point is an important characteristic of polyoxyethylene nonionic surfactant The greater the water solubility of ethylene oxide, the greater the turbidity.

(2) When the hydrophobic chains are the same, the CMC of non-ionic surfactants is lower than that of ionic surfactants.

(3) The liquid surface tension decreases with the increase of surfactant concentration.

(4) The solubilization of surfactants has nothing to do with CMC.

2. Please explain the following words in English.

(1) Surfactant.

(2) Critical micelle concentration.

(3) Surface tension.

(4) Cloud point.

Lesson 2　Surfactant Adsorption

Adsorption—the tendency of atoms or molecules to locate at a particular interface in concentrations different froth those in the surrounding bulk media—is an important process in practically all aspects of our live. The adsorbing species may be gas, solvent, or solute and the interface may be solid-solid, solid-liquid, solid-vapor, liquid-liquid, or liquid-vapor. In addition, adsorption may be termed "positive", where the concentration of adsorbed species at the interface is greater than that in the bulk or "negative" when the opposite is true.

On a rather simple basis adsorption can be divided into two main classes; ① physical adsorption or physisorption, in which the forces and processes leading to adsorption may be reversible, nonspecific, and of relatively low energy (on a molecular basis); or ② chemisorption, in which the faces and processes of adsorption are specific in nature, more irreversible, and of higher energy.

The possibilities for variations in the energy and mechanism of an adsorption process are limited only by the number of combinations of materials and circumstances one can imagine. There also exists, of course, the possibility of combined or hybrid processes that cannot be so simply classified.

The basic concept governing our understanding of the adsorption of surface active molecules at interfaces is the Gibbs adsorption isotherm, which relates the surface's excess concentration of the adsorbed species to the surface or interracial tension of the system. The surface excess may be functionally defined as the concentration of adsorbing species (i. e. , surfactant) at the interface over and above that in the bulk phase or phases. Because of its simplicity, the Gibbs approach to quantifying surfactant adsorption is open to some criticism on theoretical grounds. However, its long and useful service as a fundamental tool for understanding the phenomena involved warrant discussion.

The fundamental principle underlying our present understanding of surface activity is the Gibbs adsorption equation.

$$\Gamma_2 = - \frac{1}{RT(d\sigma/d\ln c)} \tag{1}$$

Where Γ_2 is the surface excess concentration of adsorbed species (i. e. , surfactant) above that in the bulk phase, R is the gas constant, T is the temperature (K), σ is the surface or interfacial tension, and c is the solute concentration.

In the discussion of the performance of a surfactant in lowering the surface tension of a solution it is necessary to consider two aspects of the process: ① the concentration of surfactant in the bulk phase required to produce a given surface tension reduction, and ② the maximum reduction in surface tension that can be obtained, regardless of the concentration of surfactant present.

Because the extent of reduction of the surface tension of a solution depends on the substitution of surfactant for solvent molecules tit the interface, the relative concentration of surfactantin the bulk and interfacial phases should serve as an indicator of the adsorption efficiency of a given surfactant and, therefore, as a quantitative measure of the activity of the material at the solution-vapor interface. For a homologous series of straight-chain surfactants in water, $CH_3(CH_2)_n$-S, where S is the hydrophilic head group and n is the number of methylene units in the chain, tin analysis of the thermodynamics of transfer of a surfactant molecule from the bulk phase to the interface indicates that the effectiveness of surfactant adsorption, reflected as a given surface-tension lowering, will be directly related to the length of the hydrocarbon chain, in agreement with experimental observation.

It is reasonable to expect that change in the hydrophobic character of the surfactant will produce parallel changes in adsorption efficiency, as is found for cmc values. In nitrogen-based cationic surfactants, the presence of short-chain alkyl groups (fewer than four carbon atoms) attached to the nitrogen scorns to have little effect on the efficiency of adsorption of the molecule. The dominant factor will always be the length of the primary hydrophobic chain. That effect is true whether the alkyl groups are attached to a quaternary ammonium group, an amine oxide, or a heterocyclic nucleus such as pyridine.

Within limits, the nature of the charge on an ionic surfactant will have little effect on the efficiency of surfactant adsorption. It will, again, be the nature of the hydrophobic group that predominates. Some increase in adsorption efficiency will be seen if the counterion is one that is highly ion paired. The addition of neutral electrolyte to an ionic surfactant solution will produce a similar result in increasing the efficiency of adsorption.

Polyoxyethylene (POE) nonionic surfactants with the same hydrophobic group and an average of $7 \sim 30$ oxyethylene (OE) units, exhibit adsorption efficiencies that follow an approximately linear relationship based on the free energy of transfer of $-CH_2-$ and OE groups, respectively, from the bulk phase to the interface. Available data indicate that the efficiency of adsorption decreases slightly as the number of OE units on the surfactant increases.

Although the efficiency of surfactant adsorption at the solution-vapor interface is dominated by the nature of the hydrophobic group and is relatively little affected by the hydrophilic head group, it is often found that the second characteristic of the adsorption process, the minimum surface tension obtainable (adsorption effectiveness), will be much more sensitive to other factors and will quite often not parallel the trends found for adsorption efficiency. When one discusses the effectiveness of adsorption, as defined as the maximum lowering of surface tension regardless of surfactant concentration, the value of σ_{min} is determined by the system (i. e. , the combination of temperature, solute content, surfactant characteristics, etc.) and represents a fixed point of reference. The value of σ_{mim} for a given surfactant will be determined by one of two factors: ① the solubility limit or Krafft temperature (T_k) of the compounds, or ② the critical micelle concentration. In either case, the maximum amount of surfactant adsorbed will be reached, for all practical purposes, at the maximum bulk concentration of free (i. e. , monomeric) surfactant.

Because the activity of a surfactant below T_k cannot reach its theoretical maximum as determined by the thermodynamics of surfactant aggregation, it will also be unable to achieve its maximum level of adsorption at the solution-vapor interface. It is therefore important to know the value of T_k for a given system before considering its application. Most surfactants, however, are employed well above their Krafft temperature, so that the controlling factor for the determination of their effectiveness will be the cmc.

Words and Expressions

reversible [rɪˈvɜrsəb(ə)l] *adj.* 可逆的,正反两用的

combination [ˌkɑmbɪˈneɪʃ(ə)n] *n.* 化合

chemisorption 化学吸收,化学吸附

theoretical [ˌθiəˈretɪk(ə)l] *adj.* 理论的

isotherm [kemɪˈzɔrpʃən] *n.* 等温线,恒温线

straight-chain *adj.* 直链的

tension [ˈtenʃ(ə)n] *n.* 张力,拉力,拉伸;矛盾

surface-tension *n.* 表面张力

homologous [həˈmɒləgəs] *adj.* 相应的,同源的

nitrogen-based *adj.* 氮基

effectiveness [ˌfekˈtɪvnɪs] *n.* 效率,效果,有效性

short-chain *adj.* 短链的

sensitive [ˈsensətɪv] *adj.* 敏感的,灵敏的;感光的

Notes

1. Krafft temperature:krafft 点,SDS 等离子型表面活性剂在水中的溶解度随着温度的变化而变化。当温度升高至某一点时,表面活性剂的溶解度急剧升高,该温度称为 krafft 点。

2. cmc:critical micelle concentration,临界胶束浓度,一般认为,表面活性剂在溶液中,超过一定浓度时会从单个离子或分子缔合成为胶态的聚集物,即形成胶束。溶液性质发生突变时的浓度,即胶团开始形成时溶液的浓度,称为临界胶束浓度。

Exercises

1. Please describe the difference between physical absorption and chemical absorption in surfactants in English.

2. Please search an article on surfactants and give the absorption mechanism of surfactants in the article.

Lesson 3　Wetting

The wetting of a surface by a liquid and the ultimate extent of spreading of that liquid over the surface are important aspects of practical surface chemistry. Many of the phenomenological aspects of the wetting processes have been recognized and quantified since early in the history of observation of such processes. Despite the new information developed during the last twenty-five years, a great deal remains to be learned about the mechanisms of movement of a liquid across a surface.

One of the primary characteristics of any immiscible, two- or three-phase system containing two condensed phases, at least one of which is a liquid, is the contact angle of the liquid, θ, on the second condensed phase. The contact angle of one liquid on another, while being of theoretical interest, is normally of little practical importance. Of more practical and widespread importance is the contact angle of a liquid directly on a solid. For liquids on solids, the contact angle can be viewed as a material property of the system, assuming that certain precautions are taken in the collection and interpretation of data, and that liquid absorption and surface penetration (i. e. , swelling) are taken into consideration.

When a drop of liquid is placed on a solid surface, the liquid will either spread across the surface to form a thin, approximately uniform film or it will spread to a limited extent but remain as a discrete drop on the surface. The final condition of the applied liquid on the surface is taken as an indication of the wettability of the surface by the liquid or the wetting ability of the liquid on the surface. The quantitative measure of the wetting process is taken to be the angle, θ, that the drop forms on the solid as measured through the liquid in question.

In the case of a liquid that forms a uniform film (i. e. , where $\theta=0$) , the solid is said to be completely wetted by the liquid: or the liquid wets the solid. If a finite contact angle is formed ($\theta > 0$) , some investigators describe the system as being partially wetted. If θ is 90°or more, the system is generally considered as nonwetting.

While the contact angle of a liquid on a solid may be considered a characteristic of the system, that will be true only if the angle is measured under specified equilibrium conditions-time, temperature, component purities, etc. Contact angles are very easy measurements to make (with a little practice) and can be very informative; but if the proper precautions are not taken, they can be very misleading.

The contact angle may be geometrically defined it' s the angle formed by the intersection of the two planes tangent to the liquid and solid surfaces at the perimeter of contact between the two phases and the third surrounding phase. Typically, the third phase will be air or vapor, although systems in which it is a second liquid essentially immiscible with the first are of great practical importance. The perimeter of contact among the three phases is commonly referred to as the three-phase contact line or the wetting line.

The great utility of contact angle measurements stems from their interpretation based on e-quilibrium thermodynamic considerations. As a result, most studies are conducted on essentially static systems in which the liquid drop has (presumably) been allowed to come to its final equilibrium value under controlled conditions. In many practical situations, however, it is just its important, or perhaps more important, to know how fast wetting and spreading occur as to know what the final equilibrium situation will be. That will especially be true in situations where the process in question requires that wetting bring about the displacement of one phase by the wetting liquid. Typical examples would be detergency, in which a liquid or solid soil is displaced by the wash liquid; petroleum recovery, in which the liquid petroleum is displaced by an aqueous fluid; textile processing, in which air must be displaced by a treatment solution (dyeing or waterproofing treatments, for example) in order to obtain a uniform treatment; and certain cosmetic and pharmaceutical applications. Because of the economic importance of these and other processes, the dynamic aspects of the wetting processes may be of importance.

The interpretation of data on contact angles must be donewith the understanding that the system in question has been sufficiently well controlled so that the angle measured is the "true" angle and not a reflection of some contaminant on the solid surface or in the liquid phase of interest. Contact angles, for example, can be extremely useful as a spot test of the cleanliness of sensitive surfaces such as glass or silicon wafers for microelectronics fabrication. If the surface is contaminated by something such as oil, a drop of water placed on it will have a relatively large contact angle and the contamination will be immediately apparent.

For systems which have"true" nonzero contact angles, the situation may be further complicated by the existence of contact angle hysteresis. That is, the contact angle one observes may vary depending on whether the liquid is advancing across fresh surface (the advancing contact angle, θ_A) or receding from an already wetted surface (the receding contact angle, θ_R) . As an operational convenience, many, if not most, static contact angles reported are in fact advancing angles. It is generally found that $\theta_A > \theta > \theta_R$.

Obviously, contact angle measurements and their interpretation are not without problems. However, because of the ease of making such measurements, the low cost of the necessary apparatus, and the potential utility of the concept, they should be seriously considered as a rapid diagnostic tool for any process in which wetting phenomena play a role.

There exists a variety of simple and inexpensive techniques for measuring contact angles, most of which are described in detail in various texts and publication and will only be mentioned here. The most common direct methods include the sessile drop, the captive bubble the sessile bubble, and the tilting plate. Indirect methods include tensiometry and geometric analysis of the shape of a meniscus. For solids for which the above methods are not applicable, such as powders and porous materials, methods based on capillary pressures, sedimentation rates, wetting times, imbibition rates etc. have been developed.

Words and Expressions

wetting   [ˈwetɪŋ]   *v.* 使潮湿, 把……弄湿
sedimentation   [ˌsedimenˈteɪʃ(ə)n]   *n.* 沉淀, 沉降
immiscible   [ɪˈmɪsəbl]   *adj.* 不混溶的, 不互溶的
detergency   [dɪˈtɜːdʒənsɪ]   *n.* 脱垢力
penetration   [ˌpenəˈtreɪʃn]   *n.* 浸透, 渗透
contact angle   接触角
discrete   [dɪˈskriːt]   *adj.* 离散的, 分离的
only if   只是在……的时候
tangent   [ˈtændʒənt]   *adj.* 切线的, 接触的
depend on   依靠, 依赖
dyeing   [ˈdaɪɪŋ]   *n.* 染色, 染色工艺
sessile drop   座滴法
contaminated   [kənˈtæmɪneɪtɪd]   *adj.* 被污染的
captive bubble   俘泡法
hysteresis   [ˌhɪstəˈriːsɪs]   *n.* 滞后现象(作用), 迟滞性
tilting plate   倾斜板

Exercises

1. How to distinguish whether a solid surface is wetted? Please answer this question in English.

2. Using the wettability of surfactants, what aspects can surfactants be applied to? Please answer it in English.

Reading Material: Critical Micelle Concentration

Because there are many factors that have been shown to strongly affect the critical micelle concentration, the following discussion has been divided so as to somewhat isolate the more important factors.

Any discussion of cmc data must be tempered with the knowledge that the reported values must not be taken to be absolute, but reflect certain variable factors inherent in the procedures employed in their determination. The variations in *cmc* found in the literature for nominally identical materials under supposedly identical conditions must be accepted as minor "noise" that should not significantly affect the overall picture (assuming, of course, that good experimental technique has been employed).

The length of the hydrocarbon chain in a surfactant is a major factor determining the cmc. The cmc for an homologous series of surfactants decreases logarithmically as the number of carbons in the chain increases. For straight-chain hydrocarbon surfactants of about 16 carbon atoms or less bound to a single terminal head group, the cmc is usually reduced to approximately one-half of its previous value with the addition of each methylene group. For nonionic surfactants, the effect can be much larger, with a decrease by a factor of 10 following the addition of two carbons to the chain. The insertion of phenyl and other nonhydrocarbon linking groups, branching of the alkyl group, and the presence of polar substituents on the chain can produce different effects on the cmc.

The relationship between the hydrocarbon chain length and cmc for ionic surfactants generally fits the Klevens equation

$$\log_{10}cmc = A - Bn_c \tag{1}$$

where A and B are constants specific to the homologous series under constant conditions of temperature, pressure etc. , and n_c is the number of carbon atoms in the chain. Values of A and B for a wide variety of surfactant types have been determined, and some are listed in Table 1.

Table 1   Klevens Constants for Common Surfactant Classes

Surfactant Class	Temperature/℃	A	B
Carboxylate soaps (Na⁺)	20	1.85	0.30
Carboxylate soaps (K⁺)	25	1.92	0.29
n-Alkyl- 1-sulfates (Na⁺)	45	1.42	0.30
n-Alkyl-2-sulfales (Na⁺)	55	1.28	0.27

149

Continued Table 1

Surfactant Class	Temperature/℃	A	B
n-Alkyl-l-sulfonates	40	1.59	0.29
p-n-Alkylbenzene sulfonates	55	1.68	0.29
n-Alkylammonium chlorides	25	1.25	0.27
n-Alkyltrimethylammonium bromides	25	1.72	0.30
n-Alkylpyridinium bromides	30	1.72	0.31

For more complex surfactant structures, the following generalizations serve as a good guide.

(1) Ionic surfactants having two or three neighboring ionic groups exhibit a linear relationship between cmc and chain length similar to Eq. (1), although they usually have a lower Krafft temperature and a higher cmc than the corresponding singly charged molecule of the same hydrophobic group.

(2) In surfactants having branched structures, with the head group attached at some point other than the terminal carbon, the additional carbon atoms off of the main chain contribute about one-half the effect of main chain carbons. Except for the lower members of a series, the relationship between carbon number and cmc follows a linear relationship similar to Eq. (1).

(3) For surfactants that contain two hydrophobic chains, such as the sodium dialkylsulfosuccinates, it is generally found that the cmc for the straight-chain esters follow the Klevens relationship. The cmc levels for the branched esters of equal carbon number occur at higher concentrations.

(4) In the alkylbenzene sulfonates, the aromatic ring makes a contribution equivalent to about 3.5 carbon atoms.

(5) For surfactants that contain ethylenic unsaturation in the chain, the presence of a single double bond increases the cmc by a factor of 3 ~ 4 compared to the analogous saturated compound. In addition to the electronic presence of the double bond, the isomer configuration (cis or trans) will also have an effect, with the cis isomer usually having a higher cmc.

(6) The presence of polar atoms such as oxygen or nitrogen in the hydrophobic chain (but not associated with a head group), results in an increase in the cmc. The substitution of an-OH for hydrogen, for example, reduces the effect of the carbon atoms between the substitution and the head group to half that expected in the absence of substitution. If the polar group and the head group are attached at the same carbon, that carbon atom makes little or no contribution to the hydrophobic character of the chain.

(7) A number of commercial surfactants are available in which all or most of the hydrophobic character is derived fromthe presence of polyoxypropylene groups. The observed effect of such substitution has been that each propylene oxide group is equivalent to approximately 0.4

methylene carbons.

    polyoxypropylene   *n.* 聚氧丙烯, 聚丙基醚

    trans   *n.* 反, 反式

    propyleneoxide   *n.* 氧化丙烯, 环氧丙烷

    alkylbenzene sulfonates   *n.* 烷基苯磺酸盐

    cis isomer   *n.* 顺式异构体

    dialkylsulfosuccinates   *n.* 二烷基磺基琥珀酸盐

Reading Material: Surface Tensions of Solution

1. The Nature of a Liquid Surface–Surface Tension

It is common practice to describe a liquid surface as having an elastic "skin" that cause the liquid to assume a shape of minimum surface area, its final shape being determined by the "strength" of that skin relative to other external factors such as gravity. In the absence of gravity, or when suspended in another immiscible liquid of equal density, a liquid will spontaneously assume the shape of a sphere. In order to distort the sphere, work must be done on the liquid surface, increasing the total surface area and therefore the free energy of the system. When the external force is removed, the contractile "skin" then forces the drop to return to its equilibrium shape.

While the picture of a skin like a balloon on the surface of a liquid is easy to visualize and serves a useful educational purpose, it can be quite misleading, since there is no skin or tangential force as such at the surface of a pure liquid. It is actually an imbalance of forces on surface molecules pulling into the bulk liquid and out into the adjoining fluid phase which produces the apparent skin effect. The forces involved are, of course, the same Van der Waals interactions that account for the liquid state and for most physical interactions between atoms and molecules. Because the liquid state is of higher density than the vapor, surface molecules are pulled away from the surface and into the bulk causing the surface to contract spontaneously. For that reason, it is more accurate to think of surface tension (or surface energy) in terms of the amount of work or energy required to increase the surface area of the liquid isothermally and reversibly by unit amount, rather than in terms of some tangential contractile force. Thermodynamically the surface tension, σ is defined as

$$\sigma = \Delta G / \Delta A \qquad (1)$$

where ΔG is the free energy change associated with the creation of new surface area, ΔA.

Most commonly encountered room-temperature liquids have surface tensions against air or their vapors that lie in the range of 10 ~ 80 mN/m. Water, the most important liquid commonly encountered in both laboratory and practical situations, lies at the upper scale of what are considered normal surface tensions with a value in the range of 72 ~ 73 mN/m at room temperature, while hydrocarbons reside at the lower end, falling in the lower middle 20 s.

Because of the mobility of molecules at fluid interfaces, it is not surprising to find that temperature has a large effect on the surface tension of a liquid (or the interfacial tension between two liquids). An increase in surface mobility due to an increase in temperature will clearly increase the total entropy of the surface, and thereby reduce its free energy, resulting in a negative temperature coefficient for σ.

2. Surfactants and the Reduction of Surface Tension

Since the surface tension of a liquid is determined by the excess energy of the molecules in the interfacial region, the displacement of surface solvent molecules by adsorbed solute will directly affect the measured value of σ. It is the relationship between the chemical structure of an adsorbing molecule and the rate and extent of adsorption under given circumstances that differentiates the various surfactant types and determines their utility in applications where surface tension lowering is of importance.

In aqueous solutions, the interface between the liquid and vapor phases involves interactions between relatively densely packed, highly polar water molecules, and relatively sparse, nonpolar gases. The result is a large imbalance of forces acting on the surface molecules and the observed high surface tension of water (72. 8 mN/m). If the surface solvent molecules are replaced by adsorbed surfactant molecules with lower specific excess energy, the surface tension of the solution will be decreased accordingly, with the amount of reduction being related to the surface excess concentration of solute (i. e. , the excess over the concentration in the bulk solution) and the nature of the adsorbed molecule.

If the vapor phase is replaced by a condensed phase that has a higher molecular density and more opportunity for attractive interaction between molecules in the interfacial region, the interfacial tension will be reduced significantly. In the case of water, the presence of a liquid such as octane, which the interacts only by relatively weak dispersion forces, lowers the interfacial free energy to 52 mN/m. If the extent of molecular interaction between phases can be increased by the introduction of polar groups that interact more specifically with the water, as, for instance, in octanol, the interfacial energy lowering will be greater (to 8. 5 mN/m). Clearly, any alteration in the nature of the molecules composing the liquid surface would be expected to result in a lowering of the interfacial energy of the system. And therein lies the basic explanation for the action of surfactants in lowering the surface and interfacial tension of aqueous solutions.

The same qualitative reasoning also explains why most surfactants will not affect the surface tension of organic liquids—the molecular nature of the liquid and the surfactant are not sufficiently different to make adsorption a particularly favorable process. Or if adsorption occurs, the energy gain is not sufficient to produce a significant change in the surface energy. The actions of fluorocarbon and siloxane surfactants are exceptions since the specific surface free energy of such materials may be significantly lower than that of most hydrocarbons. They will therefore be positively adsorbed at hydrocarbon surfaces and lower the surface tension of their solutions.

Words and Expressions

tangential   [tæn'dʒenʃl]   *adj.* 切向的, 切线的

siloxane   [saɪ'lɒkˌseɪn]   *n.* 硅氧烷, 有机硅氧烷

contractile   [kən'træktaɪl]   *adj.* 可收缩的, 可变窄的

fluorocarbon   [ˌflʊroʊ'kɑrbən]   *n.* 碳氟化合物

reside   [rɪ'zaɪd]   *vi.* 居住在, 定居于

octanol   ['ɒktənɒl]   *n.* 辛醇

Reading Material: Solution Properties of Surfactant

1. The Phase Spectrum of Surfactants in Aqueous System

For those concerned with the physico-chemical properties of surfactants it is important to understand the basic solution properties of such species. Most discussions of surfactants in solution are concerned with properties at low concentrations, i. e. , systems that contain what may be thought of as "simple" species such as free surfactant molecules and basic aggregates such as micelles (even though micelles may be quite complex). In many applications, however, cosmetics included, the situation may be much more complex, so that one must consider the presence and effects of higher level association structures and other interactions. The variety of possible surfactant interactions that one may encounter may be thought of as its "spectrum" in terms of the potential effects the surfactant may have in a given system.

2. Surfactant Solubility and the Krafft Temperature

The chemical nature of surfactant molecules is responsible for their unique characteristics in terms of self-association (or aggregation), adsorption at interfaces, and various possible interactions with co-solutes. However, before activity can become apparent, the system must contain a surfactant concentration sufficiently high so that the relevant characteristics of the system will be altered enough to be notable. Leaving aside the possibility of simply not introducing enough surfactant into the system, an important first aspect of the surfactant in question is its solubility.

A primary driving force for the development of synthetic surfactants in this century was the problem of the precipitation of the fatty acid soaps in the presence of multivalent cations such as calcium and magnesium. While most common surfactants have a substantial solubility in water, that characteristic can change significantly with changes in the length of the hydrophobic tail, the nature of the head group, the valency of the counter ion, and other aspect of the solution environment. For many ionic materials, for instance, it is found that water solubility increases as the temperature increases. It is often observed that the solubility, of such materials will undergo a sharp, discontinuous increase at some characteristic temperature, referred to as the Krafft temperature, T_k. Below T_k, the solubility of the surfactant is determined by the crystal lattice energy and heat of hydration of the system, its with normal crystalline materials. The concentration of the monomeric species in solution will be limited to some equilibrium value determined by those properties. Above T_k, the solubility of the surfactant monomer may increase to the point at which self-association or micelle formation begins and the associated species becomes the thermodynamically favored form. The concentration of monomeric surfactant will again be limited by the energetics of the system.

A micelle may be viewed simplistically as structurally resembling the solid crystal or a crystalline hydrate, so that the energy change in going from the crystal to the micelle will be less than the change in going to the monomeric species in solution, even though the process is usually seen as going through the monomeric solution phase. Thermodynamically, the formulation of micelles favors an overall increase in solubility. The concentration of surfactant monomer may increase or decrease slightly at higher concentrations (at a fixed temperature) , but micelles will be the predominant form of surfactant present above a certain concentration, the critical micelle concentration or cmc. The total solubility of the surfactant, then, will depend not only on the solubility of the monomeric material, but also on the "solubility" of the micelles.

The Krafft temperatures of a few common surfactant types are given in Table 1.

Table 1   The Krafft Temperature (T_k) of Typical Ionic Surfactants

Surfactant	$T_k/^\circ C$	Surfactant	$T_k/^\circ C$
$C_{12}H_{25}SO_3Na$	38	$C_{10}H_{21}CH(CH_3)C_6H_4SO_3Na$	32
$C_{14}H_{29}SO_3Na$	48	$C_{16}H_{33}CH(CH_3)C_6H_4SO_3Na$	61
$C_{16}H_{33}SO_3Na$	57	$C_{16}H_{33}OCH_2CH_2OSO_3Na$	36
$C_{12}H_{25}OSO_3Na$	16	$C_{16}H_{33}(OCH_2CH_2)_2OSO_3Na$	24
$C_{14}H_{29}OSO_3Na$	30	$C_{16}H_{33}(OCH_2CH_2)_3OSO_3Na$	19
$C_{16}H_{33}OSO_3Na$	45		

Words and Expressions

multivalent   [ˌmʌltiˈvælənt]   *adj.* 多价的, 多义的

counter ion   [ˈkaʊntər ˈaɪən]   *n.* 反荷离子, 抗衡离子

Unit 2   Surfactants

After completing this unit, you should be able to:

1. Give the definition of surfactant and describe the structure characteristic of surfactant.
2. List the four catagories of surfactant on the basis of their ionic or nonionic character.
3. List the example of surfactant in your daily life.

Lesson 1   Surfactants

The term surfactant is shorthand for the more cumbersome "surface active agent". It is common practice to depict surfactant molecules as ball and stick figures:

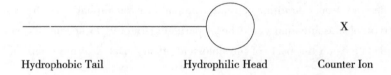

Hydrophobic Tail   Hydrophilic Head   Counter Ion

In this cartoon, the hydrophobe is represented by a stick; the ball represents the hydrophilic grouping, which may carry a positive and/or a negative charge or no charge; X represents the counter ion required for electroneutrality of the molecule.

Surfactants as a group have the ability to modify the interface between various phases. Their effects on the interface are the result of their ability to orient themselves in accordance with the polarities of the two opposing phases. Thus the polar (hydrophilic) part of the surfactant molecule can be expected to be oriented toward the more polar (hydrophilic) phase at a given interracial contact site. Similarly, the nonpolar (lipophilic) portion of the surfactant molecule should contact the nonpolar (lipophilic) phase. Each surfactant molecule has a tendency to reach across (bridge) the two phases, and such substances have, therefore, also been called amphiphilic.

One of the prerequisites for an amphiphilic molecule is possession of at least one polar and at least one essentially nopolar portion. The orientation of a 1, 2-dodecanediol molecule at a mineral-oil/water interface is readily predictable from the preceding discussion, but the positioning of 1, 12-dodecanediol at a similar interface is not as obvious: it would be expected to be different and more complex than that of the 1, 2-isomer. Despite their chemical similarity, the surfactant activities of these two compounds can be expected to be different. It is apparent from this that a surfactant's behavior or utility, e. g. , as an emulsion stabilizer, is unrelated to its empirical formula. Instead, a surfactant's spatial configuration, i. e. , the molecule's structure, plays a critical role in determining its application in cosmetics.

Classification or categorization of the thousands of different surfactants on the basis of generally recognized principles is clearly desirable. Thus it would appear practical to base such a scheme on the surfactant's functionality. Creating groupings based on such functional groups could in all likelihood be made without regard to commonly accepted chemical or physical characteristics. A typical functional scheme was developed in the CTFA (Cosmetic Ingredient Handbook) by creating six functional categories for surfactants:

Surfactants, Cleansing Agents

Surfactants, Emulsifying Agents

Surfactants, Foam Boosters

Surfactants, Hydrotropes

Surfactants, Solubilizing Agents

Surfactants, Suspending Agents

An entirely different means for classification might be based on the nature of the hydrophobic portions of surfactants. Such a classification would create groups based on the presence of hydrophobes derived from paraffinic, olefinic, aromatic, cycloaliphatic, or heterocyclic hydrophobes. This type of classification could be of particular interest to specialists who may's wish to compare substances on the basis of physiological effects related to the origin of the lipophilic constituents.

The most useful and widely accepted classification is based on the nature of the hydrophilic segment of the surfactant molecules. This classification system has universal acceptance and has been found to be practical throughout the surfactant industry. This approach creates four large groups of chemicals: amphoterics, anionics, cationics, and nonionics. This system categorizes surfactants on the basis of their ionic or nonionic character, does not consider differences in the hydrophobic (nonpolar) segment, and ignores functionality.

Words and Expressions

hydrophilic [ˌhaɪdrəˈfɪlɪk] *adj.* 亲水(性)的

stabilizer [ˈsteɪbəlaɪzər] *n.* 稳定剂

lipophilic [ˌlɪpəˈfɪlɪk] *adj.* 亲脂的, 亲油的

olefinic [ɒlɪfɪˈnɪk] *adj.* 烯的, 烯(烃)族的

prerequisites [priˈrɛkwəzəts] *n.* 先决条件, 前提

aromatic [ˌærəˈmætɪk] *adj.* 芳香的, 芳香族的

1, 2-dodecanediol *n.* 1, 2-十二烷二醇

cycloaliphatic 脂环族的, 环烯烃

1, 2-isomer *n.* 1, 2-异构体

heterocyclic   [ˌhetərəˈsaɪklɪk]   *adj.* 杂环的

Foam Boosters   [foʊm ˈbustərz]   *n.* 增泡剂

functionality   [ˌfʌŋkʃəˈnælɪti]   *n.* 官能度, 功能

Hydrotropes   *n.* 水溶助长剂

in all likelihood   十有八九, 多半

paraffinic   [ˌpærəˈfɪnɪk]   *n.* 石蜡族的, 烷(烃, 族)的

derived from   由……获得, 由……引出

shorthand   [ˈʃɔːrthænd]   *n.* 速记(法), 简写

CTFA   化妆品成分手册

Notes

CTFA：Cosmetic Toiletry and Fragrance Association, 美国化妆品、盥洗品和芳香品协会，现已改名为个人护理产品委员会—Personal Care Products Council (PCPC)。除了一些植物的提取物和某些特定原料，CTFA 化学名称与 INCI 化学名称原则上没有大的区别，因两者均是以科学性和其他拉丁文和英文名称为基础来制定化妆品成分的名称。

Exercises

1. Please list the six functional categories for surfactants in English.
2. Please describe the structure of surfactant in English.
3. Search an article on surfactant and translate the abstract of the article.

Reading Material: Utility and Selection of Surfactants in Cosmetics

Those who require and use surfactants tend to define surfactants on the basis of performance. Regardless of diverse theoretical considerations, practicing cosmetic formulators have developed a usage classification that they find practical in their day-to-day activities. As a rule, a surfactant is soluble in at least one of the contacting phases and is used to perform one or more of the following tasks: ① Clean (Detergency), ② Wet, ③ Emulsify, ④ Solubilize, ⑤ Disperse, or ⑥ Foam.

Surfactants are useful for creating a wide variety of dispersed systems, such as suspensions and emulsions. They cleanse and solubilize and are required not only during manufacture but are also essential for maintaining an acceptable level of physical stability of thermodynamically unstable systems, such as emulsions. Few modem cosmetic products exist that do not depend on one or more surfactants to create and maintain their desired characteristics.

It is the practitioner's responsibility to select one or more surfactants that can perform the task at hand. As a result of prior experience, formulators usually can identify those surfactant structures that can be expected to be most useful for achieving the desired goal.

The cosmetic formulator's choice of surfactants is more limited than that of the industrial chemist. Some of the criteria influencing selection are briefly noted below:

(1) Safety—Adverse reactions to any surfactant used in a finished cosmetic must be minimized.

(2) Odor and Color—Odoriferous or deeply colored surfactants can affect the esthetics of a finished product and should be avoided.

(3) Purity—Impurities present in some surfactants may make the surfactant unacceptable for cosmetic use.

Despite these and other limitations and the obvious requirement of cost, the cosmetic chemist must make a selection from about 2 000 different commercially available surfactants.

The selection for the specific formulation task requires insight into the general chemical characteristics of surfactants and an understanding of the physic-chemical behavior of these amphiphilcs.

The nomenclature of surfactants can become very complex and confusing. For the purpose of labeling of cosmetics in accordance with U. S. regulation, the Cosmetics, Toiletry and Fragrance Association has created names for cosmetic ingredients. It is likely that these names will soon be accepted in many other countries in the hope that a worldwide agreement on this INCI nomenclature can be reached between governmental regulatory agencies and the trade associations concerned with cosmetics.

Rules for creating these names are included in the International Cosmetic Ingredient Dictionary. The names are intended to be descriptive for laypersons as well as the more technically oriented.

Words and Expressions

esthetics [εˈsθɛtˌiks] *n.* 美学, 审美学

laypersons [ˈleˌipɜrs(ə)n] *abbr.* 常民, 外行的患者

INCI nomenclature INCI 命名法

amphiphilc [ˌæmfəˈfˌlˌik] *adj.* 两亲的, 中极两性的

Notes

1. Cosmetics, Toiletry and Fragrance Association: 化妆品、化妆品和香水协会。
2. International Cosmetic Ingredient Dictionary: 国际化妆品成分词典。

Lesson 2　Amphoterics (A)

Surfactants are classified as amphoteric if—and only if—the charge(s) on the hydrophilic, head change as a function of pH. Such surfactants must carry a positive charge at low pH and a negative charge at high pH and may form internally neutralized ionic species (zwitterions) at an intermediate pH. These features of amphoterics are illustrated below with the behavior of lauraminopropionic acid at various pH levels:

$[R-NH_2-CH_2-CH_2-COOH]^+ X^-$

Low pH: The surfactant molecule is a cation.

$[R-NH_2-CH_2-CH_2-COO]^+$

Intermediate pH: The surfactant molecule is a zwitterion.

$[R-NH-CH_2-CH_2-COO]^- C^+$

High pH: The surfactant molecule is an anion.

In this example, R represents the lauryl alkyl group, while X^- and C^+ are the required counter ions. The behavior of this substance must be compared with that of lauryl betaine:

$[R-N(CH_3)_2-CH_2-COOH]^+ X^-$

Low pH: The surfactant molecule is a cation.

$[R-N(CH_3)_2-CH_2-COO]^+$

Intermediate pH: The surfactant molecule may be a zwitterions.

$[R-N(CH_3)_2-CH_2-COO]^+$

High pH: The surfactant molecule is a cation and an anion.

Lauryl betaine contains a quaternary nitrogen atom regardless of pH. The ionization of the carboxylic acid group is, however, pH dependent, and internal compensation is possible. Lauryl betaine is properly classified as a quaternary surfactant. In cosmetic usage, betaines and related molecules exhibit some functions associated with amphoterics. Although some authorities have at times classified betaines as amphoterics, they are classified here as quaternaries.

The hydrophilic groups in amphoterics commonly are primary, secondary, or tertiary amino groups and an ionizable acidic group, i. e. , $-COO^-$, $-SO_3^-$, or rarely $-OPO_3^-$ on the same molecule. Two types of amphoterics exist: ① Alkyl(amido) betaine, ② Alkyl Substituted Amino Acids.

(1) Alkyl Betaines

The primary substances in this group are the N-alkyl derivatives of N-dimethyl glycine. The alkyl groups may include some heteroatoms. The group also includes a few hydroxypropyl sulfonates (sultaines). The betaines are manufactured by reaction of an alkyldimethylamine with chloroacetic acid. Unless carefully purified, these products may be contaminated with starting product or glycolic acid.

The betaines exhibit good water solubility. As a rule, they are compatible with all types of surfactants but may form complexes with anionics near their isoelectric pH. Some typical structures follow:

$$CH_3$$
$$|$$
$$R\text{—}N^+\text{—}CH_2COO^-$$
$$|$$
$$CH_3$$

Laural Betaine ($R = C_{12}H_{25}$)

The betaines are solids but are available primarily in aqueous solutions. They are stable and foam well. They are used primarily as hair- and skin-conditioning agents and are widely employed in shampoos. They reportedly have the ability to lower the protein-swelling tendencies and irritation potential of alkyl sulfates. They act as foam boosters and viscosity-increasing agents in shampoos. Due to their mildness and ability to lower irritation of anionics they are often used in baby shampoos.

(2) Alkylamido Betaine

These substances are synthesized by acylation of the primary amino group of aminoethyl ethanolamines ($NH_2\text{—}CH_2\text{—}CH_2\text{—}NH\text{—}CH_2\text{—}CH_2OH$) with a long chain (fatty) acid derivative. The resulting cyclic 2-alkyl hydroxyethyl imidazoline is hydrolyzed in the subsequent alkylation step with chloroacetic acid or ethylacrylate to yield a complex mixture of mono- or dicarboxy alkyl derivatives.

$$RCONHCH_2CH_2NCH_2COONa$$
$$|$$
$$CH_2CH_2OH$$

Sodium Capryloamphoacetate ($R = C_7H_{15}$)

$$RCONHCH_2CH_2NCH_2COOH$$
$$|$$
$$CH_2CH_2OCH_2CH_2COOH$$

Lauroamphodipropionic Acid ($R = C_{11}H_{23}$)

Alkylation with, for example, hydroxypropylsulfonic acid, yields a more complex tertiary amine. Commercial products are mixtures containing soaps and the hydrolysis product of the alkylating agent. They are sold as salts (usually sodium) or as free acid. At or near neutral pH they may exist in zwitterionic form. The amide linkage in these molecules may be subject to hydrolysis, but no report of chemical instability in cosmetics has been published.

Alkylamido alkyl amines are generally water soluble and are compatible with most other cosmetically useful surfactants. They reportedly reduce the tendency of anionics to elicit eye irritation without significantly interfering with their foaming characteristics.

These amphoterics exhibit substantivity to hair and skin proteins and act as conditioning and antistatic agents. Their primary use is in shampoos and miscellaneous skin cleansers. They are, however, not widely used as detersive surfactants (cleansing agents) and are not effective emulsifying agents.

Words and Expressions

amphoteric   [ˌæmfəˈterɪk]   *adj.* 有酸碱两性的

irritation   [ˌɪrɪˈteɪʃ(ə)n]   *n.* 激怒, 刺激, 兴奋

zwitterions   兼性离子, 两性离子

mildness   [maɪldnəs]   *n.* 温和

betaine   [ˈbiːtəːn]   *n.* 甜菜碱

imidazoline   [ˌɪmɪdəˈlin]   *n.* 咪唑啉, 间二氮杂环戊烯

ionization   [ˌaɪənaɪˈzeɪʃ(ə)n]   *n.* 离子化, 电离

isoelectric   [ˌaɪsoʊˈlektrɪk]   *adj.* 等电位的

derivative   [dɪˈrɪvətɪv]   *n.* 衍生物, 派生的事物

acylation   [ˈæsəleʃən]   *n.* 酰化作用

hydrolysis product   *水解产物*

alkylating agent   烷基化剂

antistatic agents   抗静电剂

be subject to   受支配, 可以……的

as a rule   通常

associated with   与……相联系, 与……有关

regardless of   不管

compared with   与……比较

Notes

两性离子(英语:zwitterion)是总电荷为0,电中性的化合物,但是带正电和负电的原子不同。有些化学家将此术语限定为未具有相邻正负电荷的化合物。此定义将诸如氧化胺的化合物排除。

Exercises

1. Please list the zwitterionic surfactants in your daily used cosmetics and describe the effect of them.

2. Search the English literatures about zwitterionic surfactants and describe the development trend of zwitterionic surfactants.

Reading Material: Amphoterics (B)

Alkyl substituted amino acids are prepared by alkylation of various synthetic and natural amino acids or by the addition of an amine to an α, β unsaturated alkanoic acid. Some typical structures follow:

$$R-N\begin{array}{l}CH_2CH_2COO^- \\ \\ CH_2CH_2COOH\end{array} \quad Na^+$$

Sodium Lauriminodipropionate ($R = C_{12}H_{25}$)

$$RNH-CH_2-CH_2-COOH$$

Myristaminopropionic Acid ($R = C_{14}H_{29}$)

As a group, these compounds exhibit excellent stability under conditions of cosmetic use.

Alkyl substituted amino acid foam copiously, especially above their isoelectric point. At low pH levels they behave as cationics and foam poorl. They can be used as emulsifiers. As amphoterics, they are substantive to hair and find their most important uses in various hair coloring and hair conditioning products.

Words and Expressions

Myristaminopropionic acid　肉豆蔻丙酸
isoelectric　*adj.* 等电位的，零电位差的
Sodium Lauriminodipropionate　月桂酰胺二丙酸钠
alkanoic acid　*n.* 链烷酸，烷酸
Alkyl substituted amino acids　烷基取代氨基酸
unsaturated　[ən'sætʃəˌreɪtɪd]　*adj.* 不(未)饱和的

Lesson 3　Anionics (A)

All surfactants in which the hydrophilic head of the molecule carries a negative charge are classified as anionics. The group of anionic surfactants includes types of great industrial importance and substances widely used in cosmetics. As a rule, they are inactivated or even form complex precipitates in the presence of cationic surfactants. This complexation is generally attributed to salt formation in which the ionized species react in stoichiometric proportions. The complexes may be solubilized in aqueous systems containing large amounts of anionics.

For the sake of classification, anionic surfactants may be subdivided into five major chemical classes and subgroups:

(1) Acylated Amino Acids and Acyl Peptides
(2) Carboxylic Acids (and Salts)
　(a) Alkanoic Acids
　(b) Ester-functional Carboxylic Acids
　(c) Ether-functional Carboxylic Acids

(3) Sulfonic Acid Derivatives

    (a) Taurates

    (b) Isethionates

    (c) Alkylaryl Sulfonates

    (d) Olefin Sulfonales

    (e) Sulfosuccinates

    (f) Miscellanous Sulfonates

(4) Sulfuric Acid Derivatives

    (a) Alkyl Sulfates

    (b) Alkyl EtherSulfates

(5) Phosphoric Acid Derivatives

The members of these five classes form water soluble salts with alkali metals and low molecular weight amines, especially alkanol amines.

The members of subgroups (1) and (2) above depend on ionization of the carboxylic acid group for aqueous solubility. On the other hand, salts formed with alkaline earths or heavy metals exhibit limited or no solubility in water.

Alkanoic Acids. The most important members of this subgroup are the fatty acids derived from plant and animal glycerides. These natural acids normally possess an even number of carbon atoms and carry only one carboxylic acid group. The unsaturation in natural fatty acid is almost exclusively cis. A few natural fatty acids also contain a hydroxy group, in addition, some alkanoic acids are prepared synthetically, especially those in which the alkyl group is branched (iso).

Fatty acids are obtained by the alkaline hydrolysis of fats and oils. Acidification after removal of unsaponifiables yields a water insoluble fatty acid blend named on the basis of its source, e. g. , olive oil fatty acids. Specific fatty acids (oleic acid), can be isolated from these mixtures by various chemical and physical techniques.

Alkanoic acids, as a group, are important industrial chemicals and are used in the synthesis of many types of substances. One of the most important modifications of alkanoic acids is reduction to fatty alcohols, which are then processed further to yield a variety of surfactants. Free alkanoic acids are of limited use in cosmetics, but the water soluble salts (soaps) are amongst the most useful surfactants known. Soaps have been utilized as cleansers and detersive agents since antiquity. In modern practice, soaps are the alkali or low molecular weight amine salts of alkanoic acids. Their water solubility depends on the pH of the system and on the cation. As a rule, potassium salts are more soluble than the sodium salts. The alkanoic acids are weak acids, with a reported pK_a of about 5 ~ 6. Therefore soaps—as salts of weak acids—yield alkaline aqueous solutions due to their dissociation in water.

The solubility of alkali or amine salts of alkanoic acid in water decreases as the length of the alkyl chain increases. Thus, sodium stearate, especially in the presence of some free stearic acid, is insolubleenough to permit manufacture into soap bars. The alkaline earth and metal salts of alkanoic acids are water insoluble. Thus, calcium salts precipitate in aqueous systems leading to the formation of so-called soap scum.

Alkanoic acid salts in which the alkyl chain contains about ten or fewer cations are not useful as surfactants, i. e. , they do not foam well, have no detersive qualities, and are poor e-mulsifiers. The stearic acid of commerce contains about 45% of octadecanoic and 55% of hexa-decanoic acids. The product may include small amounts of oleic acid and other acids normally found in the starting lipid. Modem grades of stearic acid are primarily prepared by hydrogenation of soybean fatty acids. For illustrative purpose, the following structures are included:

$$[RCOO^-]_2Mg^{2+}$$

Magnesium stearate ($R = C_{17}H_{35}$, $C_{15}H_{33}$)

$$[C_8H_{17}CH\!=\!CHC_7H_{14}COO]^-[NH(CH_2CH_2OH)_3]^+$$

TEA oleate

Water soluble soaps are used as skin and hair cleaning agents, while the insoluble derivatives (e. g. , zinc laurate or magnesiun stearate) are used for lubricating solids to improve flow properties, act as binders, and increase the viscosity of nonaqueous systems. Sodium stearate is soluble in warm ethanol and tends to gel upon cooling. Thus this substance has found extensive use in the formulation of alcohol-based stick deodorants.

Water soluble and water insoluble soaps are good emulsifiers, the former primarily for o/w emulsions, while soaps such as aluminum stearate tends to form w/o emulsions. As a role, oleic acid salts are especially useful emulsifiers, but their usage is restricted by the tendency of this unsaturated acid to form malodorous or discolored peroxidation products.

One of the most important applications of soaps is represented by shaving soaps in general. Regardless of the method of shaving (brush, brushless, or aerosol) , soap stocks from various sources are commonly blended to provide the shaver with copious and rapidly generated foam that lasts until shaving is completed.

The topical use of soaps for skin cleansing is considered safe, although it has been shown that soaps can elicit adverse reactions on skin during closed patch testing.

Words and Expressions

complexation　[kɒmplek'seɪʃ(ə)n]　*n.* 络合,络合作用

lubricating　['lubrɪˌkeɪt]　*adj.* 润滑的,润滑

stoichiometric　[stɔɪkɪˈɒmɪtrɪk]　*adj.* 化学计量的

deodorant　[diˈoʊdərənt]　*adj.* 除臭的; *n.* 除臭剂

solubilize　['sɒljəbəˌlaɪz]　*v.* (使)增溶,溶剂化

aerosol ['erə̩sɑl] *n.* 气溶胶, 雾化器

peptides ['peptaɪd] *n.* 缩氨酸, 肽

malodorous [mæl'oʊdərəs] *adj.* 有恶臭的, 恶臭的

alkaline ['ælkə̩laɪn] *adj.* 碱的, 氨基醇类

attributed to 归因于

glyceride ['glɪsə̩raɪd] *n.* 甘油酯, 脂肪酸丙酯

stoichiometric proportions 化学计量比

alkaline hydrolysis 加碱水解

for the sake of 为了……

hexadecanoic acids 十六烷酸, 棕榈酸

alkali metals 碱金属

patch testing 斑片试验, 皮肤过敏试验

Notes

patch testing:斑贴试验。由试验致敏物质接触皮肤后引起的表皮过敏反应(如接触性皮炎、湿疹等)。将被试物配成适当浓度,用 1 cm^2 纱布四层浸湿后放在皮肤上,覆以玻璃纸,再以胶布固定,在 24 ~48 h 后观察其反应。

Exercises

1. Please list the anionic surfactants in your daily used cosmetic and describe the effect of them.

2. Search the English literatures about anionic surfactants and describe the purpose of anionic surfactants.

Reading Material: Anionics (B)

1. Acylated Amino Acids and Acyl Peptides

These substances are usually prepared by the reaction of a natural amino acid or of a peptide with a long-chain fatty acid derivative. In this reaction, primary amino groups are converted into acylated amido groups. This destroys the zwitterionic character of the amino acid or of the peptide and increases the acidity of the carboxylic acids. After completion of the acylation, these acid groups, are frequently, neutralized with a suitable alkali. The following examples illustrate some of the structures:

$$[RCON(CH_3)_2CH_2COO]^-TEA^+$$

TEA lauroyl sarcosinate ($R = C_{11}H_{23}$)

$$[HOOCCH_2CH_2CHCOO]^-Na^+$$
$$|$$
$$RCHO\!-\!NH$$

Sodium stearoyl glutamate($R = C_{17}H_{35}$)

Collagen or some of its hydrolysis products are the most common sources of the protein. The level of hydrolysis (enzymatic or chemical) is not generally specified, and so-called acylated peptides are likely to contain considerable amounts of acylated aminoacids. Since some of the amino acids contain more than one site for acylation (e. g. , hydroxyproline), the end products are probably rather complex mixtures and may include some simple soaps.

The acyl sarcosinates (derived from N-methyl glycine) occupy a special niche in cosmetics. These substances behave like soaps. The key to their performance and mildness is the fact that the carboxyl group has a lower pKa than that of typical fatty acids. The salts of the sarcosinates are water soluble and can be used at pH levels near or even slightly below neutrality.

Acylated amino acids, depending on molecular weight and complexity, foam modestly and are generally viewed as exceptionally mild. They find use in skin and hair cleansing products and have been included in syndet bars. They reportedly exhibit substantivity to hair and skin proteins. Members of this class are sometimes identified as amphoteric. Under conditions of cosmetic usage (pH 4 to 9), acylated amino acids or peptides carry an anionic charge that is neutralized by a suitable cation. Their reported substantivity to hair or skin is the result of some unidentified protein—protein interaction unrelated to the charge on the surfactant's head group.

Acylated amino acids are amides and subject to chemical (or enzymatic) hydrolysis. They are, however, stable at the pH commonly found in cosmetics but are subject to microbial attack. Preservation against spoilage remains a major problem, especially in the case of the peptide-derived products.

2. Ester-functional Carboxylic Acids

One type of ester-functiional carboxylic acid is the small group of esters derived from polycarboxylic acids in which at least one of the carboxylate groups is free to form a sail. A typical example is stearyl citrate, the monoester of stearyl alcohol with citric acid.

An entirely different type is represented by the acylation compounds of lactyl lactate. In their synthesis, two molecules of lactic acid are believed to react with each other, and the dimer then reacts with a fatty acid. The structure of a typical emulsifier created by this reaction is shown below:

$$[RCOOCH（CH_3）COOCH（CH_3）COO]^-Na^+$$

Sodium StearoyI Lactylatc（$R=C_{17}H_{35}$）

Compounds belonging to this class are safe for use in foods（baked goods）, are occasionally used as cosmetic emulsifiers, and are reported to condition hair and skin.

3. Ether-functional Carboxylic Acids

Compounds belonging to the group of ether-functional carboxylic acids have recently gained some prominence in cosmetic usage. They may be viewed its alkylethers of polyethyleneglycol in which the terminal OH group has been oxidized to a carboxy group. The principal synthetic route depends on the alkylation（e. g., with chloroacetic acid）of an ethoxylated alcohol（D. 3. a）. As derivatives of glycolic acid, their pKa is quite low. The presence of the polymeric ether group increases the water solubilily of these substances even if the starting alcoholic hydrophobe is relatively bulky. A typical structure is provided below for illustrative purposes:

$$[R（OCH_2CH_2）_5OCH_2COO]^-Na^+$$

Sodium Trideceth-Carboxylate（$R=C_{13}H_{27}$）

The water solubility of the free acids increases with increasing levels of ethoxylation. In this form, these compounds are useful as emulsifiers. Neutralization（usually with sodium ion）yields surfactants with detersive and solubilizing properties. These compounds are stable under normal conditions of cosmetic use. Compounds of this type have been shown to reduce the skin irritation potential of other anionic surfactants and are generally milder themselves.

Words and Expressions

syndet   n. 合成皂,合成洗涤剂,合成清洁剂

Sodium Stearoyl Lactylatc   n. 硬脂酰乳酸钠

sarcosinates   n. 肌氨酸盐

baked   [beikt]   v.（在烤炉里）烘烤,焙

substantivity   n. 直接性,亲和性

ethoxylation   n. 乙氧基化,乙氧基化物

acylated   adj. 酰化的,酰基化

detersive   [di:'tərsiv]   adj. 有清洁效力的; n. 清洁剂

stearyl citrate   n. 硬脂酰柠檬酸酯

Lesson 4 Cationics（A）

The cationic surfactants carry a positively charged nitrogen atom on the hydrophobe. The positive charge may be permanent, i. e., independent of pH, as in the true quaternaries, or may be pH dependent, as in amines. Cationics can be further subdivided as follows:

(1) Quaternaries

    (a) Alkyl Benzyl Dimethylammonium Salts

    (b) Tetraalkylammonium Salts

    (c) Heterocyclic Ammonium Salts

(2) Alkyl Amines

(3) Alkyl Imidazolines

This important group of cosmetic surfactants is distinguished from alkyl amines and amphoterics by the fact that all quaternaries carry a tetrasubstituted N-atom. Quaternaries can elicit toxic and allergic responses. They are customarily used at love levels (for antimicrobial or conditioning effects), and documented reports of adverse reactions are relatively rare. Quaternaries are substantive to proteins and their tendency to penetrate stratum corneum is limited.

(1) Alkyl Benzyl Dimethylammonium Salts

This group of quaternaries is derived from aliphatic tertiary amines carrying at least two methyl groups by reaction with a benzyl halide. The nature of the alkyl groups on the N-atom is variable and may include chains carrying other heteroatoms. A typical illustrative structure follows:

$$[RCOOCH(CH_3)COOCH(CH_3)COO]^-Na^+$$

Cetalkonium Chloride ($R = C_{16}H_{33}$)

These substances are solids but are available as solutions or suspensions. They are excellent hair-conditioning agents, and some are useful antimicrobial agents, especially benzalkonium chloride. In this compound, the broad spectrum antimicrobial activity depends on the length of the alkyl group, with the C_{10} to C_{14} chains being the most effective. The benzalkonium salts are also used as emulsifiers and suspending agents

(2) Tetraalkylammonium Salts

This group of quaternaries differs from the other quaternaries in the fact that none of the four substituent groups on the N-atom is specified. As a result, these compounds include widely varying members exhibiting different solubilities and physical properties.

The substituent groups on the N-atom may be identical or may include one or two polyoxyethylene or polyoxypropylene chains. The solubilities are dependent on the characteristics of the substituent groups. The tetraalkyl ammonium salts—in common with other cationics—exhibit substantivity to skin and hair proteins and are generally incompatible with anionic surfactants. Under conditions of cosmetic use, these substances are stable. Some illustrative example follow:

$[RN(CH_3)_3]^+CH_3SO_3^-$ Cetrimonium Methosulfate ($R = C_{16}H_{33}$)

$\{[R(OCH_2CH_2)_4]_2N(CH_3)_2\}^+Cl^-$ Dilaureth-4 Dimonium Chloride ($R = C_{12}H_{25}$)

(3) Heterocyclic Ammonium Salts

Substances in this group result from the alkylation heterocyclic N-containing amines with a suitable alkylhalide. The number of surfactants in this class is limited; most of them are derived from pyridine, morpholine, isoquinoline, or imidazoline.

Except for the imidaozoline (imidonium) derivatives, these quaternaries are stable under conditionsof cosmetic use. Almost all of them are water soluble solids. Some representative structures follow:

$$[R(OCH_2CH_2)_5OCH_2COO]^-Na^+$$

$$\left[R-N\bigcirc\right]^+ Cl^-$$

$$\left[R\overset{N}{\underset{N}{\diamond}}N\overset{CH_2CH_2OCH_2CH_2COOH}{\underset{CH_2CH_3}{}}\right]^+ CH_3CH_2OSO_3^-$$

The heterocyclic quaternaries include an important group of antimicrobial agents, e. g., cetylpyridinium chloride and dequalinium chloride, a bisquaternary. Most of tile heterocyclic ammonium salts find use as hair and skin conditioning agents.

Words and Expressions

cationic　[ˌkæt'aˌənˌk]　*adj.* 阳离子性的, 阳离子的

antimicrobial　*n.* 抗菌剂, 杀菌剂; *adj.* 抗菌的

permanent　['pɜrmənənt]　*adj.* 永久的, 持久的

spectrum　['spektrəm]　*n.* 光谱, 型谱, 领域, 系列

quaternaries　['kwɑtərˌnɛriz]　*n.* 季铵盐, 季铵盐类

suspending　[sə'spendˌŋ]　*v.* 悬浮, 悬浊, 悬置; *adj.* 悬吊的, 悬浮的

allergic　[ə'lɜːrdʒˌk]　*adj.* 变态反应的, 对……过敏的

tetrasubstituted　*adj.* 四元取代的

conditioning　[kən'dˌʃnˌŋ]　*n.* 保持(头发或皮肤等的)健康, 养护

heteroatoms　杂原子

distinguished from　不同于……

antimicrobial activity　抗微生物(抗菌)活性

stratum　['streˌtəm]　corneum 角质层

in common with　和……一样

benzalkonium　*n.* 苄烷铵, 苯甲烃铵

dequalinium　*n.* 地喹氯铵

cetylpyridinium　*n.* 十六烷基吡啶

Exercises

1. Please list the cationic surfactants in your daily used cosmetics and describe the effect of them.

2. Search the English literatures about cationic surfactants and describe the development trend of cationic surfactants.

Reading Material: Cationics (B)

(1) Alkyl Amines

Long chain alkyl amines, whether primary, secondary, or tertiary, are hydrophobic. They act as surfactants only after they have been neutralized, usually with a strong inorganic or organic acid. The free amines can be made more hydrophilic by, forming an amido amine from an acid chloride and an aliphatic diamine. Hydrophilicity of the free amine can be further enhanced by treating a primary or secondary amine with ethylene oxide, which attaches a polyoxyethylene chain to the amino-N.

The alkyl amines are waxy solids of variable water solubility. They are stable under conditions of cosmetic usage. Structures of some typical examples follow:

$$R-N\begin{cases} CH_3 \\ CH_3 \end{cases}$$

Dimethyl Pahnitamine ($R = C_{16}H_{33}$)

$$R-N\begin{cases} CH_2CH_2OH \\ CH_2CH_2OH \end{cases}$$

Dihydroxyethyl Tallowamine (R = Tallow alkyl)

$$R-N\begin{cases} (CH_2CH_2O)_xH \\ (CH_2CH_2O)_yH \end{cases}$$

PEG 50 Stearamine ($R = C_{17}H_{35}$; $x + y = 50$)

Ethoxylation creates amines that are sometimes compatible with anionics. Unethoxylated amines are more basic and thus generally not compatible with anionics.

Neutralized alkyl amines are positively charged and are used as substantive hair and skin conditioning agents. As a group the alkylamines are useful emulsifying and dispersing agents.

(2) Alkyl Imidazoline

The heterocyclic alkyl imidazolines are the precursors of tile alkylamido alkylamines and of the quaternized imidonium derivatives. The alkyl imidazolines are available as aqueous solutions. They are not resistant to hydrolysis tinder adverse pH conditions. The structure of a typical example is shown below:

$$R \overset{\displaystyle N-CH_2CH_2OH}{\underset{\displaystyle N}{\vphantom{x}}}$$

Behenyl Hydroxyethyl imidazoline ($R = C_{21}H_{43}$)

Alkyl imidazolines can be used as emulsifiers and conditioning agents. Their use in cosmetic formulation is limited by their questionable stability and the need for neutralization.

Words and Expressions

Unethoxylated   *adj.* 未乙氧基化

aliphatic diamine   *n.* 脂肪族二胺

Alkyl imidazolines   *n.* 烷基咪唑啉

Dihydroxyethyl Tallowamine   *n.* 二羟乙基牛油胺

Lesson 5　Nonionics (A)

Nonionic surfactants are substances in which the molecule carries no charge at the pH levels of cosmetic use. The hydrophobe can be highly variable, but the hydrophilic head generally includes a polyether group or at least one OH group. For the sake of this discussion, nonionics are subdivided into five large groups: ①Ethers, ②Esters, ③Alkanolamides, ④Amine Oxides.

1. Ethers

Ethers as a group are widely used in cosmetic and pharmaceutical products because of their good resistance to hydrolytic reactions. The ethers of interest contain not only a (repeated) —C—O—C—grouping but also a terminal C—O—H grouping.

For the sake of convenience, some ethers derived from naturally occurring lipids are also included in this group. The ether group is subdivided as follows:

(a) Ethoxylated Alcohols;

(b) Ethoxylated Lanolin and Castor Oil;

(c) Ethoxylated Polysiloxanes;

(d) Alkyl Glucosides;

(e) POE/PPG Ethers.

· Ethoxylated Alcohols

In the synthesis of these ethers, a hydrophobic alcohol, a sterol, or a phenol is treated with ethylene oxide. The alcohols are those already described in the synthesis of alkyl sulfates and those useful as secondary emulsifiers. In addition, these alcohols are sometimes modified by reacting them with propylene oxide before ethoxylation. The sterols are those found in nature (cholesterol and its reduction product, soybean sterols, and other phytosterols). Alkyl phenols (octyl from diisobutylene and nonyl from tripropylene) are the starting materials for useful nonionic surfactants.

The ethoxylation is carried out under pressure and in the presence of alkaline catalysts. Usually ethoxylated alcohols yield a mixture of POE ethers of varying levels of ethoxylation, free alcohols are common contaminants. The n values are—at best—an approximation of the average number of ether-forming ethylene oxide units. A myristyl alcohol ethoxide identified as Myreth-8 commonly contains no more than about 15% to 20% of the 8-mole POE derivative of the starting alcohol. The remaining substances show a Gaussian distribution of EO units with a peak of about 8.

The water solubility is a function of the degree and distribution of ethoxylation. The HLB of these ethers can range from about 3 to 18. They may be liquid or pasty solids. They foam poorly and are used primarily as o/w, or w/o emulsifiers or solubilizers. Most of them are considered innocuous. Three typical structures follow to illustrate the chemistry of the alkoxylated alcohols:

$R(OCH_2CH_2)_nOH$

Myreth 8 ($R = C_{14}H_{29}; n = 8$)

$RC_6H_4(OCH_2CH_2)_nOH$

Octoxynol 10 ($R = C_8H_{17}; n = 10$)

$R(OCH(CH_3)CH_2)_m(OCH_2CH_2)_nOH$

PPG 25-Laureth-5 ($R = C_{12}H_{25}; m = 25; n = 5$)

2. Esters

Ester are among the most frequently used surfactant in cosmetics. Esters are subject to hydrolysis, but the pH conditions for such reactions in cosmetics do not prevail. In addition, nonionic esters are among the safest surfactants available to formulators and are common constituents of processed foods. For the sake of facilitating the discussion of specific groups, nonionic ester-type surfactants are divided its follows:

(a) Ethoxylated Glycerides;

(b) Glycol Esters;

(c) Monoglycerides;

(d) Polyglyceryl Esters;

(e) Carbohydrate Derived Esters;

(f) Ethoxylated Carboxylic Acids;

(g) Sorbitan Esters;

(h) Trialkyl Phosphates.

· Ethoxylated Glycerides

Substances belonging to this class of compounds are derived from mono-, di-, or triacyl glycerides. Ethoxylation of an α-monoglyceride leads primarily to alkoxylation of the —OH group of the glyceride, although some β-ethoxylation is possible. The structure of these derivatives probably conforms to the following:

$$RCOOCH_2CHCH_2(OCH_2CH_2)_xOH$$
$$|$$
$$(OCH_2CH_2)_yOH$$

PEG-20 Glyceryl Oleate ($R = C_{17}H_{33}$; $x+y = 20$)

Diacylglycerides can also be ethoxylated to yield compounds of the following type:

$$RCOOCH_2CHCH_2OOCR$$
$$|$$
$$(OCH_2CH_2)_nOH$$

PEG-12 Glyceyl Dioleate ($R = C_{17}H_{33}$; $n = 12$)

Finally, triacyl glycerides undergo ethoxylation with some transesterification to yield complex mixtures that may include some ethoxylated carboxylic acids and some more complex esters resembling those derived from sorbitan. One of many potential structures is shown below:

$$RCOOCH_2$$
$$|$$
$$CH^-(OCH_2CH_2)_5OOR$$
$$|$$
$$R''COOCH_2$$

PEG-5 Hydrogenated Corn Glycerides RCO, R'CO, and R''CO = acyl radicals

When reactions of this type are carried out on glycerides containing Oil-acids (e. g. , castor oil) ethoxylation of —OH groups may occur.

The water solubility, of ethoxylated glycerides, depends on the degree of ethoxylation, which can be quite high, e. g, PEG-200 castor oil. These substances are stable to hydrolytic reactions at the pH levels normally encountered in cosmetic practice. They are used as emulsifiers, suspending agents, and solubilizers.

Words and Expressions

sterol   [ˈsterɔul]   *n.* 固醇, 甾酮, 甾醇, 甾醇类

sorbitan   山梨糖醇酐, 脱水山梨糖醇, 山梨聚糖

phytosterol   [ˌfaiˈtɒsterɔul]   *n.* 植物甾醇, 植物固醇

transesterification   [trænsəsterəfiˈkeiʃən]   *n.* 酯交换

phenol   [ˈfiˌnɔul]   *n.* 苯酚, 石炭酸, 酚

myristyl   [ˈmiˌriˌstiˌl]   alcohol *n.* 肉豆蔻醇, 十四烷醇

173

propylene oxide   *n.* 环氧丙烷

Gaussian   [ˈgaʊsɪən]   distribution *n.* 高斯分布

average number   *n.* 平均数

Exercises

1. Please list thenonionic surfactants in your daily used cosmetics and describe the effect of them.

2. Search the English literatures aboutnonionic surfactants and describe the development trend of nonionic surfactants.

Reading Material: Nonionics (B)

1. Alkanolamides

Alkanolamides are the acylation products of various alkanolamines. Two types of alkanolamides exist. One of these, the superamides, are prepared from the 1 : 1 mole ratio of the amine and the acylating species, yielding primarily water insoluble N-acylalkanolamide. This type of product is contaminated with esteramine and probably the ester amide in which both the OH- and NH- functions are acylated. When two moles of an alkanol amine are reacted with one mole of the acylating species, the so-called Kritchevsky condensates are formed. They may contain all of the components identified in the description of the 1 : 1 product. In addition, these water soluble condensates may contain alkanolamine soaps and derivatives of morpholine and piperazine. The INCI nonlenclature does not differentiate between these two types, both of which are available commercially.

A third group of compounds results when an acid amide is allowed to react with ethylene oxide. In this case, water solubility is determined by the degree of ethoxylalion.

Some representative structures, describing the predominant components provided in the INCI Dictionary, follow:

$RCON(CH_2CH_2OH)_2$

Palmitamide DEA ($R = C_{15}H_{31}$)

$RCONH(CH_2CH_2O)_nH$

PEG-12 Cocamide ($RCO = Coco$ acyl: $n = 12$)

The 1 : 1 and the Kritchevsky condensates find their primary uses as foam boosters and foam stabilizers in shampoos. They are only rarely used as emulsifiers. The ethoxylated amides are relatively stable to hydrolysis and find use as emulsifiers at low pH levels (e. g. , in antiperspirants) .

2. Amine Oxides

Amine oxides are formed from tertiary aliphatic amines by oxidation, generally with hydrogen peroxides. The tertiary amine may be a straight chain or part of a heterocyclic system. There has been some claim that amine oxides can be protonated at pH levels lower than those occurring in cosmetic practice. Amine oxides are generally contaminated with unreacted amines, which may account for some of the cationic behaviors of amine oxides. Pending further evidence, it seems advisable to classify amine oxides as nonionic surfactants. A typical structure is shown below:

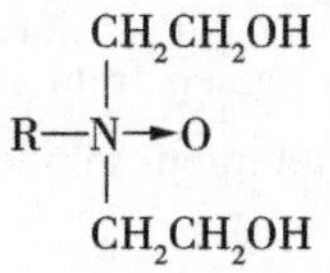

$$CH_2CH_2OH$$
$$|$$
$$R-N \rightarrow O$$
$$|$$
$$CH_2CH_2OH$$

Dihydroxyethyl Stearamine Oxide ($R = C_{18}H_{37}$)

Amine oxides are water soluble and foam well. They are used as foam boosters in shampoos and as lime soap dispersants. Amine oxides are used in hair-coloring products and reportedly can reduce the skin irritant characteristics of anionic surfactants.

3. Alkyl Glucosides

This interesting group of surfactants is prepared by reaction of hydrophobic alcohols with glucose. During the ether formation, some oligosaccharide is formed, and the reaction products could be described as tile monoalkyl ethers of a polyglycoside exhibiting an average degree of polymerization of 1.4. The structure is represented as follows:

Decyl Glucoside ($R = C_{10}H_{23}$; $n = 0 \sim 3$)

These substances foam well and arc used ill skin and hair cleansing products. The glucosides are acetals and may exhibit poor stability at low pH levels. They are reported to be mild on skin and to lower the skin irritation potential of alkyl sulfates.

Words and Expressions

Kritchevsky condensates   *n.* 克里切夫斯基凝聚体

polyglycoside   *n.* 聚苷

antiperspirants   [ˌænti'pɜrsp(ə)rənt]   *n.* 止汗药, 止汗剂

acetals   ['eˌsɪtəlz]   *n.* 乙缩醛, 缩醛类

oligosaccharide   [ˌɒlɪɡou'sækəˌraɪd]   *n.* 低聚糖

glucosides   ['ɡlukəˌsaɪdz]   *n.* 葡糖苷类, 糖苷

Lesson 6   A Brief Description of the Sulphonation Processes Used for the Anionics

1. Introduction

Sulphonation plants are scattered around the globe in units with production capacities varying between 3 000 and 50 000 tons anionic surfactants annually. Assuming an average production capacity of, say, 5 000 tons per year, there are at least 800 operational sulphonation plants around the world.

A variety of sulphonation reagents can be used for the sulphonation reaction: SO_3/air from sulphur burning and subsequent conversion of the SO_2/air formed; SO_3/air from stabilised liquid SO_3 or SO_3 stripped from 65% oleum with dried process air, 20% oleum and chlorosulphonic acid.

There are four reasons why SO_3/air raised from sulphur is becoming the predominant sulphonating agent for the manufacture of detergent actives.

1.1 *Versatility*

All kinds of organic feedstocks, like alkylbenzenes, primary alcohols, alcohol ethers, alpha-olefins and fatty acid methyl esters, can be successfully converted with SO_3/air as the sulphonating agent to high-quality sulphonate/sulphate active detergents.

1.2 *Safety*

Liquid SO_3, 65% oleum, 20% oleum and concentrated sulphuric acid are hazardous chemicals in transport, handling and storage. The EEC directive on major accident hazards is proposing that if more than 25 tons of SO_3 or its equivalent in oleum are stored on site, all regulations outlined in Annexes of Council directive 82/501/EEC have to be obeyed. Among other requirements, the following points are mandatory:

It must be demonstrated that the sulphonation is performed safely including providing details of operator training etc.

Companies need to provide "notification" to the "member state" competent authorities (in the UK, for example, the "Health and Safety Executive").

Note particularly that those to be informed include "any person outside the establishment liable to be affected by a major accident...", who should be appropriately informed of the safety measures to be taken and of the correct behaviour to be adopted in the event of an accident. In other words, if more than 25 tons of SO_3 are stored on site, the neighbours of the factory have to be informed via the local authorities.

Sulphur, either in liquid or solid form, is a least hazardous option as a starting material for the production of SO_3.

1.3 *Costs*

Sulphur is considerably cheaper as a starting material from which to raise SO_3 than liquid SO_3, 65% oleum and 20% oleum. It is also more economical than the three liquid options in transport, handling and storage. Most important of all is the problem of spent acids, resulting from sulphonation with 20% oleum (a dark, 80% strength sulphuric acid, difficult to recycle in the chemical industry) or 65% oleum (a dark, concentrated sulphuric acid, in most countries returnable to producers only at very low prices, sometimes not even compensating for transport costs) .

Stabilised liquid SO_3 requires very precise temperature control and its residues, which contain noxious stabilised residues after evaporation, and are difficult to handle and dispose of.

1.4 *Availability*

Liquid SO_3, 65% and 20% oleum and even sulphuric acid are not manufactured in all regions of the globe. In many parts of Africa and Asia sulphuric acid and different derived oleum qualities are not available. Even when sulphuric acid is produced, oleum of various strengths may not be manufactured.

2. Sulphur-based SO_3/Air Sulphonation

Fig. 1 illustrates the overall block diagram. To raise SO_3/air with a volumetric content between 4% and 7% SO_3, the process air should be dried to prevent the formation of sulphuric acid mist. Filtered ambient air is compressed (about 0. 6 bar gauge), chilled (about + 5 ℃) to remove the major part of water by condensation and subsequently dried with a desiccant (silica gel, Alumina) to arrive at a process air dewpoint of about −60 ℃.

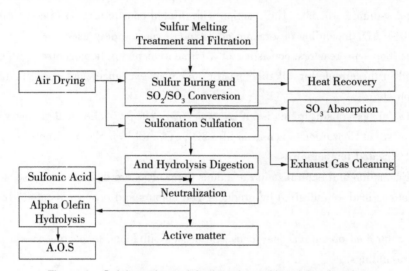

Figure 1   Sulphonation-sulphation plant with sulphur burning

Molten sulphur (150 ℃) from storage is pumped to the sulphur furnace where sulphur is converted with an excess of oxygen from the process air to SO_2(4% ~7% SO_2 by volume in "air"). The furnace outlet temperature of the SO_2/air varies with the percentage of SO_2 in air between 600 ℃ and 700 ℃, indicating the strong exothermic character of the reaction. The SO_2/air flow is cooled in an indirect air cooler from 600 ~700 ℃ to about 420 ℃.

SO_2 is converted to SO_3 in the so-called converter tower filled with 4 packed beds of V_2O_5 catalyst on a silica carrier. The reaction is highly exothermic and intermediate cooling of the process gas flow between the various beds with indirect air coolers is required.

Notwithstanding the low process air dewpoint, some sulphuric acid/oleum mist condenses in the coolers following the converter tower at temperatures of about 45 ~ 50 ℃. This highly reactive mist can affect the quality of the subsequent sulphonation reaction and therefore a high-efficiency demister is installed before the actual sulphonation step. The sulphonation reaction between SO_3 and organic feedstock is almost instantaneous and the reaction is highly exothermic.

Falling film reactors of different design are nowadays widely used for the sulphonation reaction.

After the reactor, the SO_3 exhausted gas is separated from the organic acid. The exhaust gas, containing small amounts of non-converted SO_2, unreacted SO_3 and some entrained organic acid, has to be cleaned before emission to ambient atmosphere. The organic aerosol and fine SO_3/H_2SO_4 droplets are separated from the exhaust gas flow in an electrostatic precipitator (ESP) and the gaseous SO_2 and traces of SO_3 gas are washed from the process air in a scrubber by dilute caustic solution, thus producing a mixed sulphite/sulphate solution.

The neutralization reactions can be carried out with many alkaline chemicals like caustic, ammonia and sodium carbonate. The reaction with diluted caustic to a paste containing between 40% and 70% AD, depending on organic acid type, is most widely used.

Various loop-type reactors, consisting of a circulation pump, homogeniser (where the acid is introduced in the circulating alkaline paste) and heat exchanger, are used for the complex neutralization step.

The $SO_3/$air gas-raising plant is an example of a "heavy" chemical industry operation. Highly corrosive and hazardous chemicals like SO_2, SO_3 and $SO_3/$oleum sulphuric acid mist are produced at elevated temperatures.

Thesulphonation and neutralization reactions themselves are delicate in the sense that inaccurate operations lead to undesired by-products, bad colours and poor yields of converted organic feedstocks.

Poorly controlled operations may cause hazardous situations to people on site and in the adjacent surroundings.

Words and Expressions

sulphonation [ˌsʌlfəˈneɪˌʃən] n. 磺化(作用)

notwithstanding [ˌnɑtwɪðˈstændɪŋ] prep. 虽然, 尽管

feedstock [ˈfiːdstɒk] n. 原材料, 给料

neutralization [ˌnjuːtrəlaɪˈzeɪʃ(ə) n] n. 中和

oleum [ˈəʊlˌəm] n. 发烟硫酸

falling film reactor 降膜反应器

spent acid 废酸

air cooler 空气冷却器

process air   工艺空气

electrostatic precipitator   静电除尘器

Exercises

1. Please describe the sulfonation process in combination with your internship experience.

2. Which process do you think is more important in your internship factory? Why?

Unit 1   Introduction to Fine Chemicals

After completing this unit, you should be able to:
1. Tell the difference between the processes in fine chemicals manufacture and processes for the manufacture of commodity chemicals.
2. Describe the character of fine chemicals.
3. List the examples of fine chemicals.

Lesson 1   General Introduction of Fine Chemicals

Fine chemicals are products of high and well-defined purity, which are manufactured in relatively small amounts and sold at relatively high price. Although a question of taste, reasonable limits would be 10 kton/year and $ 10/kg. Fine chemicals can be divided in two basic groups: those that are used as intermediates for other products, and those that by their nature have a specific activity and are used based on their performance characteristics. Performance chemicals are used as active ingredients or additives in formulations, and as aids in processing.

Fine chemicals form a group of products of large variety: their number exceeds 10 000. The size of the global fine chemicals market in 1993 was estimated at $ 42 000 million. The average annual growth in the period 1989–1995 was about 4. 5%. Figure 1 below shows the division of fine chemicals production by outlet.

Only 10 firms account for 75% of agrochemicals sales, while the 15 largest drug companies have a market share of only 33%. About 85% of fine chemicals are manufactured by companies of the "triad": the United States (28%), Western Europe (39%), and Japan (17%). Italy, with 4. 0 million liters reactor capacity and 71 manufacturers, topped the European fine chemicals industry. Recently India, China, and Eastern-Central European countries have gained a significant proportion of the market, as a result of the lower direct labor costs and the more relaxed environmental and safety standards. It is fair to state that the high quality of chemists in these countries has also contributed to this development. In 1993, the cost of producing fine chemicals

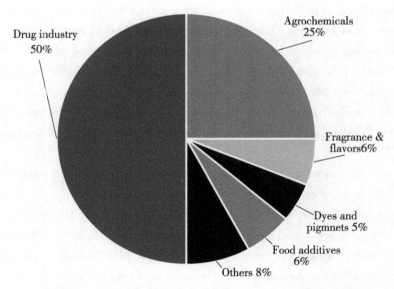

Figure 1   The division of fine chemicals production by outlet

in India was 12% below that in Europe.

Production of pharmaceuticals is heavily concentrated in four countries, namely the USA, UK, Switzerland, and Germany. Many companies active in fine chemicals business are rather small. Of course, there is no general role for an optimal size. According to Stinson sales of companies producing fine chemicals should be at least $ 50 million per year to be able to afford the costs of quality control, environmental efforts and to run the plant profitably. This translates as follows: the minimum long-term economic size of a European fine chemicals producer would be about 85 to 90 people total, who would support a reactor capacity of about 120 000 liters. However, when these firms expand to, say, more than 500 people and annual sales of over $ 100 million, they might become too inflexible to respond to changes in customer and market needs, so it is not surprising that most compounds have sales below $ 10 million per year and form a network with several competing companies.

Custom synthesis can be defined as the dedicated production for a single client, often using technology provided by the customers themselves. Custom chemicals world-wide are a $ 6 000 million-per-year-business. According to the estimate about 40% of the world fine chemicals market is produced through contract manufacturing, half of which goes to the drugs industry. Custom manufacturing develops, among others, due to inflexibility of the bigger companies. However, experts are divided in their opinions about contract manufacturing: many of them see large companies with years of experience to be winners in contract manufacturing, while others insist that room for small specialists, particularly in the life sciences market, still exists. Mullin indicates that some of the fastest-growing contract manufacturers are large firms. For example, fine chemicals at DSM grew from 2% in 1987 to 25% of the company business in 1998. As is the trend in the Chemical Industry, this growth was achieved mainly through acquisitions: Andeno, Gist-Brocades, and businesses from Bristol Meyers Squibb and Chemie Linz.

Many companies specialize in the production of chemicals grouped in "chemical trees" characterized by the same chemical roots (compounds) or the same/similar method of manufacturing. Examples are the Lonza trees based upon: ① hydrogen cyanide, ② ketene ($H_2C \! = \! C \! = \! O$) and diketene (4-methyleneoxetan-2-one), and ③ nitrogen heterocycles. Wacker Chemie has developed its chemical tree leading to acetoacetates, other acylacetates, and 2-ketones.

Processes in fine chemicals manufacture differ from processes for the manufacture of commodity chemicals in many respects.

(1) A significant proportion of the fine chemicals are complex, multifunctional large molecules. These molecules are labile, unstable at elevated temperature, and sensitive towards (occasionally even minor) changes in their environment (e. g. pH). Therefore, processes are needed with inherent protective measures (e. g. chemical or physical quenching) or a precise control system to operate exactly within the allowable range. Otherwise the yield of the desired product can drop to nearly zero.

(2) Fine chemicals are high-added-value products. In general, expensive raw materials are processed to obtain fine chemicals, and therefore, the degree of their utilization is very important. With complex reaction pathways, selectivity is the key problem to make the process profitable. Selectivity is significant also because of difficulties in isolation and purification of the desired product from many side products, especially those with physicochemical properties similar to those of the desired product (close boiling points, optical isomers, etc.). Furthermore, a low selectivity results in large streams of pollutants to be treated before they can be disposed of. Selectivity is even more an issue because in contrast with bulk chemicals production, where a limited single-pass conversion coupled with separation and recycling of unreacted raw materials is often applied, usually complete conversion is aimed at. Selectivity can be controlled by chemical factors such as chemical route, solvent, catalyst and operating conditions, but it is also strongly dependent on engineering solutions. Catalysis is the key to increasing the selectivity.

(3) In the manufacture of fine chemicals many hazardous chemicals are used, such as highly flammable solvents, cyanides, phosgene, halogens, volatile amines, isocyanates and phosphorous compounds. The use of hazardous and toxic chemicals produces severe problems associated with safety and effluent disposal. Moreover, fine chemistry reactions are predominantly carried out in batch stirred-tank reactors characterized by (i) a large inventory of dangerous chemicals, and (ii) a limited possibility to transfer the generated heat to the surroundings. Therefore, the risk of thermal runaways, explosions, and emissions of pollutants to the surroundings is greater than in bulk (usually continuous) production. That is why much attention must be paid to safety, health hazards, and waste disposal during development, scale-up, and operation of the process.

Fine chemicals are often manufactured in multistep conventional syntheses, which results in a high consumption of raw materials and, consequently, large amounts of by-products and wastes. On average, the consumption of raw materials in the bulk chemicals business is about 1 kg/kg of product. This figure in fine chemistry is much greater, and can reach up to 100 kg/kg for pharmaceuticals. The high raw materials-to-product ratio in fine chemistry justifies extensive search for selective catalysts. Use of effective catalysts would result in a decrease of reactant consumption and waste production, and the simultaneous reduction of the number of steps in the synthesis.

(4) One of the most important features of fine chemicals manufacture is the great variety of products, with new products permanently emerging. Therefore, significant fluctuations in the demand exist for a variety of chemicals. If each product would be manufactured using a plant dedicated to the particular process, the investment and labor costs would be enormous. In combination with the ever changing demand and given the fact that plants are usually run below their design capacity, this would make the manufacturing costs very high. Therefore, only larger volume fine chemicals or compounds obtained in a specific way or of extremely high purity are produced in dedicated plants. Most of the fine chemicals, however, are manufactured in multipurpose or multiproduct plants (MPPs). They consist of versatile equipment for reaction, separation and purification, storage, effluent treatment, solvent recovery, and equipment for utilities. By changing the connections between the units and careful cleaning of the equipment to be used in the next campaign, one can adapt the plant to the intended process. The investment and labor costs are significantly lower for MPPs than for dedicated plants, while the flexibility necessary to meet changing demands is provided. The need for versatility of equipment originates from the great number of products in rather limited quantities to be manufactured in the plant every year. Such versatile equipment is suitable for all the processes, although it is certainly not optimal. The most versatile reactors are stirred-tank reactors operated in batch or semibatch mode, so such reactors are mainly used in multiproduct plants. Continuous plants with reactors of small volume are sometimes used despite the small capacity required. This is the case when the residence time of reactants in the reaction zone must be short or when too much hazardous compounds could accumulate in the reaction zone.

From the foregoing it will be clear that in *fine chemicals process development* the strategy differs profoundly from that in the bulk chemical industry. The major steps are (i) adaptation of procedures to constraints imposed by the existing facilities with some necessary equipment additions, or (ii) choice of appropriate equipment and determination of procedures for a newly built plant, in such a way that procedures in both cases guarantee the profitable, competitive, and safe operation of a plant.

(5) The accuracy of analytical methods has increased enormously in the past decades and this has enabled detection of even almost negligible traces of impurities. The consequence is that both regulations and specifications for intermediates and final fine chemicals have become

stricter. Therefore, very pure compounds must often be produced with impurities at ppm or ppb level. The production of complex molecules in many cases results in mixtures containing isomers, including optical isomers. The demand for enantiomeric materials is growing at the expense of their racemic counterparts, driven primarily by the pharmaceutical industry. Therefore, both stereoselective synthesis and effective, often non-conventional methods of of racemate resolution, stimulates the development of new stereoselective catalysts, including biocatalysts. In this respect, biotechnology is becoming a competitor to classical chemical technology.

All specialists involved in the business of fine chemistry should be aware of the characteristic features of fine chemicals and their manufacture. Obviously, the most important for the development of new products is chemistry, i. e. the choice of the most appropriate route to the desired product. The role of molecular modelling is growing, particularly in the field of performance chemicals. Chemistry is also crucial in developing or improving processes for the manufacture of fine chemicals. The chemistry of fine chemicals has been dealt with in many books and review papers. Therefore, only those chemical problems arising during process development will be emphasized that can contribute to solving problems of selectivity, environmental protection, safety, and optimization when commercializing the process, especially in existing multi-product plants.

The implications of the features of fine chemicals manufacture mentioned above, are, however, not that obvious to all parties taking part in process development. On the one hand, chemical and process engineers are dedicated mainly to the engineering part in process development, often neglecting process chemistry, as this might be erroneously considered less important for a full-scale plant. On the other hand, synthetic chemists often finish their work with laboratory recipes neglecting needs of procedures for process development. The reasons for this approach have been listed by Laird: ① research chemists are interested in the properties of the product, and how it is made on a small scale is of less importance. ② the synthetic route may be long (and therefore expensive). ③ the raw materials and reagents may not be available in bulk quantities. ④ the routes may not be stereospecific—in research the required isomer will often be isolated by using a chromatographic step at the end. ⑤ the reaction conditions may involve processes which are difficult to operate in standard plants—e. g. high pressure, very low temperatures. ⑥ toxic reagents and intermediates, which are easy to handle on a gram scale, may prohibit working on a kilogram scale. ⑦ exothermic reactions, which can be handled safely in the laboratory, are more difficult to control in a full-scale plant. ⑧ effluents—easy to dispose of in the laboratory—are a serious consideration in process development. Heavy metal reagents, chlorinated solvents, toxic intermediates, etc. may have to be eliminated from the process before effluents can be safely treated. This subject is becoming very sensitive in Europe.

Words and Expressions

negative potential   *n.* 负电位

optical brighteners   *n.* 荧光增白剂

foam inhibitors   *n.* 消泡剂

corrosion inhibitors   *n.* 缓蚀剂

complex formers   *n.* 络合剂

hollow spheres   *n.* 空心球

ion-exchange resins   *n.* 离子交换树脂

low-density   *adj.* 密度低的

negative potential   *n.* 负电位

static electricity   *n.* 静电

diketene   *n.* 双烯酮

phosgene   ['fɑːzdʒiːn]   *n.* 光气, 碳酰氯

agrochemicals   *n.* 农药, 农用化学品

halogens   ['hælədʒən]   *n.* 卤素类, 卤素

triad   ['traɪæd]   *n.* 三人组合, 三位一体

isocyanates   *n.* 异氰酸酯, 异氰酸酯类

hydrogen cyanide   ['haɪdrədʒən'saɪənaɪd]   *n.* 氰化氢

multipurpose or multiproduct plants   *n.* 多用途或多产品厂

ketene   *n.* 乙烯酮

enantiomeric   [æntiə'merik]   *adj.* 对映体的

nitrogen heterocycles   *n.* 含氮杂环

racemate   ['ræsəmeɪt]   *n.* 外消旋体

acetoacetates   *n.* 乙酰乙酸酯, 乙酰乙酸盐

stereoselective   *adj.* 立体选择性, 立体构象选择性

acylacetates   *n.* 酰基乙酸盐

ppm (part per million)   *abbr.* 百万分之一

2-ketones   *n.* 2-酮

ppb (part per billion)   *abbr.* 十亿分之一

Exercises

Search the English literatures about fine chemicals and describe the development trend of fine chemicals.

Reading Material: Chemical Reaction Engineering Aspects of the Fine Chemicals Manufacture

Fine chemicals manufacture means many different things to different people, but most will agree that the common elements are; the products are complex organic molecules, with molecular weights in the hundreds, manufactured at relatively low annual rates, kilograms to perhaps a thousand tonnes. In order to understand the role of reaction engineering in fine chemicals process development and manufacture, it is important to understand the nature of the chemistry and materials involved, the development time scale and product cost structures, reactor selection and design, etc.

1. The Nature and Impact of the Chemistry

Reaction engineering the commercial viability and the selection of the best synthesis route and the development of a safe and successful manufacturing process, minimising any adverse impact on the environment and human health. The implications for the reaction engineer are clear: ① there will be very many potential by-product reactions, generating many impurities; ② the relative amounts of impurities will depend on the inherent reactivity of the molecules as well as the reactor selection and detailed design; the best process will be the result of a collaboration between the synthetic chemist and the reaction engineer; ③ there will be many options to synthesis the molecule from a variety of building blocks; ④ no single stage can be considered in isolation; ⑤ the feasibility and commercial viability of a complete chemical route may be determined by the feasibility, selectivity, yield or ease of operation of any single stage.

2. Development Costs and Time Scales

The major impact of the reaction engineering on the overall process development is at the route selection stage, ensuring that the best overall route is selected to be carried forward into the final stages of development and process design. The various types of business in the fine chemicals sector will have different development processes and time scales. The reaction engineering and chemical development will be fully integrated to ensure that the reaction conditions, reaction and extraction solvents, reactor selection and overall process design will achieve the required performance at manufacturing scale. Eventually, the engineers will go on to provide the design data for the manufacturing plant or plants. As the process firms up, it will become most important to design for the detailed mixing, mass and heat transfer necessary to achieve the target selectivity in the full-scale reactor.

3. The Batch Agitated Vessel—a Cheap and Flexible Friend

In much of the fine chemicals industry a superficial look at the reactors will give the impression that they are all very similar, predominantly batch agitated vessels in the $5 \sim 20 \text{ m}^3$ range.

It is worth considering why so many reactors are operated in batch or more commonly fed-batch mode. The traditional view from outside of the industry is that conservatism and chemists dominate and there is, therefore, resistance to adopting other types of equipment, particularly as an agitated reactor can be operated like and can look like a laboratory reaction flask. However, in reality, although in some sectors, particularly pharmaceuticals, regulation tends to favour traditional approaches, the truth is that often an agitated vessel is the best choice, and will have been selected as a result of in depth consideration of alternatives, by the engineers and chemists working together in the development team. Sometimes a fed batch reactor is the best choice or even the only choice which can be made to work and sometimes other designs are more appropriate. Examples of reactor selection on the grounds of safety, reaction selectivity and cost are described in the following section.

The major advantages of batch reactors for products of this type are cost, ability to cope with uncertainty, flexibility and ability to control.

(1) In cost terms, for the relatively low production rates required in fine chemicals manufacture, the trade off between continuous operation which requires more but smaller plant items, and batch operation requiring fewer but larger items usually shows that batch is cheaper.

(2) Considering the ability to cope with uncertainty, one advantage of the batch reactor and its associated ancillary equipment is that in many cases it is relatively easy and cheap to over-design. An additional uncertainty is how different the actual manufacturing process will be when the reactor is finally installed and commissioned compared to the process at the outset of development and at the time the reactor design is fixed. With a good grasp of the fundamentals and sound judgement, a batch reactor system can be designed to cope with a wide divergence of the process parameters. In this way, the time required to complete the detailed design, procurement and installation of the plant can be used to develop and optimize the process, reducing the development time as compared to a totally consecutive procedure by typically 18 months.

(3) Flexibility is a similar consideration to robustness. A continuous reaction stage will be designed for a very specific set of physical parameters, for example physical properties and reaction kinetics; whereas although a batch agitated vessel has some limitations, it can cope with a very wide range of materials and is generally very easy and cheap to modify to handle new conditions. The only really difficult variable to change can be pressure, as a vessel will have a specific maximum operating pressure which cannot be exceeded.

(4) The ability to control reaction selectivity in fine chemicals manufacture is very closely linked to the nature of the materials and the chemistry. There are many issues associated with accurate feeding of solids and pastes, particularly when the composition may be variable, as it is very difficult to control the solid/liquid ratio in a paste, and representative sampling for analysis is difficult. However, the most obvious problem, and one where there must be potential for new opportunities, is in on- and at-line chemical analysis. The advantage of batch reactor technology in this case is the ability to define "safe" holding positions, where the operation can be tempo-

rarily held in a state where further reaction and by-product formation is negligible, providing time to analyse and resolve problems, avoiding wasting or having to reprocess valuable materials. Holding will increase the cycle time. At the worst, this will increase the reactor size for a given design rate, which in most cases will have a negligible overall effect on process economics.

Exercises

Search the English literatures about chemical reaction in the manufacturing of fine chemicals and list the possible by-products.

Unit 2   Detergency

After completing this unit, you should be able to:
1. List the fundamental phenomena of detergency.
2. Tell the preparation process of oil-in-water cosmetic emulsions.

Lesson 1   Fundamental Phenomena in Detergency (A)

Detergency is easily the most important area of application for surfactants. Cleaning and washing are very complex processes, which even today are not fully understood. Since there is a huge variety of types of dirt and of substrates, there is no single mechanism for the process of detergency, but rather many depending on the substrate—textile, glass, metal, china.

The process of detergency can be defined in general as the removal of liquid or solid dirt from a solid, the substrate, with the aid of a liquid, the cleaning bath. Dirt itself is defined, rather simplistically, as "material in the wrong place".

In the old days, soap was the usual cleaning material, or detergent, but it has the well-known drawbacks that it forms insoluble, inactive fatty acids in an acid environment and insoluble precipitates with Ca^{2+} and Mg^{2+} in hard water. Additives such as Na_2CO_3, phosphates, etc. can prevent these disadvantages. In the last 50 years, soap has been partly replaced by synthetic detergents which do not suffer from these negative effects. The most important of these are the alkyl sulfates, alkyl aryl sulfonates and nonionic poly (ethylene oxide) derivative.

We will limit this discussion to processes using aqueous cleaning media. The following examples serve to illustrate the most important facet of the nature of the substrate:

(a) processes for the removal of dirt (liquid and/or solid) from smooth surfaces, for example the degreasing of metal, glass, ceramics (including washing dishes) , the cleaning of painted surfaces;

(b) processes for the removal of dirt from porous or fibrous materials, for example the washing of raw wool and cotton, removal of spinning oils from spun fibers, degreasing of leather, laundry, desizing of textiles.

Each of these examples represents a complex process. It is not surprising that the individual processes take very different courses; for example, emulsification is a necessary stage m the removal of wool fat from the fibers of raw wool, but is totally unsuitable for the removal of a film of oil from a polished surface. We must therefore start by considering the fundamental phenomena of detergency, which always occur, and only alter this shall we look at the subsidiary phenomena that have a role to play only in particular cleaning processes.

1. Solubilization

Solubilization in tenside micelles is important for the removal of small quantities of oily soiling from substrates. Whether from hard or textile surfaces, this removal becomes significant only once the cmc has been exceeded for nonionic tensides, and indeed even for some anionic tensides with a low cmc. Maximum effectiveness is reached only at a multiple of the cmc. The degree of solubilization of the oily dirt depends on the chemical structure of the tenside, on its concentration in the bath, and on the temperature.

At low tenside concentrations, the dirt is solubilized into more or less spherical micelles. Only a small amount of oil can be solubilized in this manner. The high tenside concentration permits large quantities of greasy dirt to be held in solution.

For ionic surfactants, the concentration used is not much larger than the cmc. Solubilization is therefore seldom sufficient to remove all oily dirt. For nonionic surfactants, the amount solubilized depends principally on the temperature of the bath, relative to the cloud point of the tenside. Solubilization of grease increases sharply close to the cloud point.

There are various views concerning soiling removed by micelles. Since the most successful detergents form micelles, some hold the opinion that micelles must be directlyinvolved in the detergency, solubilizing grease. However, the detergent action is dependent on the concentration of unassociated tenside and is practically unaffected by the presence of micelle, since only monomeric tenside molecules adsorb at the interfaces. The micelles therefore act at best, as a reservoir from which nonassociated tensides adsorbed from solution are replenished. It seems that the molecular properties of tensides with good detergency also favor the formation of micelles, so if anything these should be seen as competition rather than as an aid to dirt removal.

2. Emulsification

For emulsification, the interfacial tension between the oil droplets and the cleaning bath must be very low, so that emulsification may occur with very little mechanical work. Adsorption of the tenside at the dirt bath interface therefore plays a significant role. The suitability of the cleaning bath for emulsification of the oily dirt is insufficient to prevent all redeposition of soiling on the substrate. If emulsified oil droplets collide with the substrate. Some will stick and adopt the equilibrium contact angle. This contrasts with solubilization, in which the oily dirt is completely removed from the substrate.

Dispersion of the dirt particles in the cleaning hath alone does not amount to effective cleaning. There is no correlation between the detergency and the dispersing power of a cleaning hath. Tensides that are excellent dispersants are often had detergents, and vice versa. On the other hand, in the case of anionic and nonionic tensides there is a correlation between increased adsorption on the substrate and on dirt: furthermore, in the case of nonionic tensides, there is also a correlation between the solubilization of greasy dirt and the detergent effect.

Finally, cationic tensides have little detergent effect, since most soiling and most substrates are negatively charged in aqueous media at neutral or alkaline pit. Adsorption of the positively charged tenside ions on substrate and dirt reduces their negative electrical potential, binders the removal of dirt, and encourages its redeposition.

Words and Expressions

mechanism　[məˈkænˌks]　*n.* 物理过程, 历程, 机制

spun　[spʌn]　*n.* 氨纶, 棉纱

drawback　[ˈdrɔːbæk]　缺点, 障碍

emulsification　[ˌˌmʌlsəfəˈkeiˌʃən]　*n.* 乳化, 乳化作用

precipitate　使沉淀(出), 淀析, 析出

solubilization　[sɒljubˌlaiˌzeˌʃən]　*n.* 溶解, 增溶作用

alky laryl sulfonate　烷基芳基磺酸盐

dispersion　[diˈspɜːrʒ(ə)n]　*n.* 分散, 散布, 消散

fibrous　[ˈfaibrəs]　*adj.* 含纤维的, 纤维性的

cloud point　浊点

Exercises

1. Give a few of examples of the detergent in our daily life and list the main components in those detergent.

2. Search the English literatures about detergent and describe the development trend of detergent.

Reading Material: Fundamental Phenomena in Detergency (B)

As has already been mentioned, the processes of washing are complex. Various chemical and physical processes occur together. Both water-soluble soiling and, for example, adsorbates of pigments, grease, carbohydrates proteins, and natural and synthetic dyes must be removed from the surface.

To remove insoluble particulate soiling, the adhesion energy must be overcome, so mechanical energy must be supplied in the first step of the washing process; this is assisted by adsorption of tenside. At this stage the composition of the washing liquid controls the complete lifting of the dirt particle in the next step.

The process of lifting dirt is the reverse of flocculation, and depends on electrostatic interactions. Therefore, adsorption of ionogenic tensides or multiply charged ions such as alkali metal phosphates or silicates (builders) must be employed to ensure that substrate and soil have the same charge sign, so that their separation is electrostatically favored.

After lifting, dirt is present in the washing liquid in dispersed form, and the adsorbed tenside layers should delay or prevent the flocculation of solid particles and the coalescence of emulsified droplets. Macromolecular additives such as carboxymethylcellulose have a similar effect by steric hindrance, preventing flocculation and coalescence, which would otherwise result in redeposition of the dirt and graying of the laundry. Macromolecular additives thus also act as graying inhibitors.

Many forms of soiling have to be chemically changed before they can be removed, whether by redox processes by means of bleaching agents (combined with bleach activators and stabilizers), for example the natural dyes present in tea, wine, and fruit juice, or by enzymatic degradation of denatured protein adsorbates. Complex formation and ion exchange are other important processes.

Greasy soiling is mostly liquid at wash temperatures greater than 40 ℃ and spreads over the substrate surface in more or less unbroken layers. These oily layers need to be emulsified during the washing process; the "roll-up" already mentioned is important here.

Words and Expressions

flocculation   *n.* 絮凝, 絮结产物
coalescence   [ˌkouəˈlɛsəns]   *n.* 合并, 联合, 接合
carboxymethylcellulose   *n.* 羧甲基纤维素
redeposition   *n.* 再沉积, 再沉淀

Reading Material: Emulsions

1. Introduction

Most cooks are aware that if a mixture of oiland vinegar is shaken, a temporary suspension of the two liquid phases is formed. Separation of this salad dressing occurs in a very short time. Cooks also know that if egg yolk, oil, and vinegar are combined in the proper manner, a smooth, creamy mayonnaise is produced. The egg yolk supplies an ingredient that holds the two immiscible liquids together. In more technical terms, the egg yolk acts as an emulsifier which holds the mayonnaise emulsion together.

Emulsifiers are compounds that possess both hydrophilic and lipophilic moieties in their chemical structures. Emulsifier molecules are orientated at oil-water (or air-water) boundaries as a result of the amphiphilic nature of their chemical structures. In this orientation, the nonpolar portion of the molecule is absorbed into the nonpolar (oil) phase, and the hydrophilic portion of the molecule seeks the water phase. Due to the molecular alignment and absorption of the emulsifier at the interface of an oil-water system, the interfacial tension is reduced. The surface activity (the effect on interfacial tension at phase boundaries) of an emulsifier is determined by the polarity and solubility properties of moieties in its chemical structure.

When an emulsion is formed, the surface area between the dispersed and continuous phase increases to a great extent. The increase in surface area increases the free energy and therefore the instability of the emulsion. An emulsifier absorbing at the interface will lower the surface free energy and enhance the formation of an emulsion.

The type of emulsifier used determines the manner in which it is absorbed at the surface of the oil-water interface. Not only does this surface active agent reduce the surface tension between the two phases but it controls the type of emulsion formed (oil-water, or water-oil) and the stability of the dispersed particles.

The absorbed surfactant also tends to stabilize the emulsion by formation of electrostatic and physical barriers around the dispersed particles. Emulsion stability isinfluenced by many factors, such as phase density and mutual solubility, surface viscosity and related properties, and solvated layers around the particles, as well as by the charged electric atmosphere surrounding the dispersed particles. The addition of salts, change in pH, the addition of a nonelectrolyte or a change in temperature all effect the stability of emulsions. In any given emulsion system one emulsifier cannot be substituted for another with the same HLB without changing the emulsion stability. These changes may influence the solubility properties at the interface, the electrical atmosphere surrounding the particle, and the structure of solvated layers surrounding the dispersed phase.

Although much has been learned about emulsion systems over the years, a good method for predicting how to make a stable emulsion has not been developed. When two new phases must be emulsified, the proper emulsifier(s), concentration of emulsifier, and emulsification technique must be selected. Experience is the best guide. The HLB system is often used in selecting surfactants. This empirical method, developed by Griffin, is a useful guide in the initial stages of formulation. Ultimately, the system must be fine-tuned by trial and error.

The discussion above is concerned with the familiar types of emulsions. These are the ordinary milky, creamy, opaque emulsions-milk, ice cream, margarine, cosmetic lotions, and mayonnaise, to name a few. These are termed macroemulsions since they contain relatively large globules of one liquid phase suspended in another. The suspended phases will coalesce if permitted to remain quiescent for a prolonged period. These emulsions are all thermodynamically unstable and the attractive forces between the dispersed particles will eventually lead to agglomeration. Despite of this tendency toward coalescence, we know that useful emulsions can be formulated. Such formulations can provide a product with a shelf life in excess of normal requirements.

The microemulsions are a distinctly different type of suspension of oil and water. These are optically isotropic transparent dispersions of oil and water in combination with surfactants. In contrast to the macroemulsions, the microemulsions are thermodynamically stable; they form spontaneously when the components are brought together at the proper ratio and at the proper temperature. They may be considered single-phase solutions, or as suggested by Holt, swollen micellar solutions. The dispersed phase consists of small droplets with a size range from approximately 10 nm to 100 nm. True microemulsions form spontaneously to a state of minimum free energy of the single-phase solution. On the other hand, macroemulsions, regardless of the size of the droplets of the dispersed phase, require energy input for formation. The energy input can be mechanical agitation or homogenization, or heat, and is often a combination of the two.

Microemulsions are of great interest in many areas of application. They show a greater penetration of porous materials than do themacromulsions, allow more uniform dispersion of active substances soluble only in the dispersed phased, and yield high gloss and high film integrity in the case of waxes and paints. Moreover, microemulsions will not break down under repeated freeze-thaw cycles.

2. Preparation (Take Oil-in-Water Cosmetic Emulsions for Example)

Every cosmetics laboratory worker knows that emulsions are relatively stable mixtures of oils, fats and water and are made by mixing oil-soluble and water-soluble substances together in the presence of an emulsifying agent. Emulsions-creams and lotions form a very important part of the cosmetics market.

In general oil and aqueous phase are heated separately to 70 °C. (Occasionally temperatures up to 90 °C are recommended but lower temperature is usually preferable because the basic materials are less affected.) The phases are then mixed. Many formulations require the oil phase to be added to the aqueous phase: this method often yields very good results, but the dispersion is even better if the aqueous phase is stirred into the oil phase, which results in an unstable W/O emulsion that inverts to an O/W emulsion after further stirring. The dispersion of the oil phase is made even finer by homogenizing the emulsion. It is not necessary to stir O/W creams until they are cool; in most cases stirring may be discontinued at approximately 50 °C. The stability of the emulsion depends less on the viscosity of the external phase, and this in turn not so much on the temperature. With stearate creams it is occasionally undesirable to continue agitation below 45 °C because in these circumstances consistency and appearance of the emulsion is conditioned by the crystal form of the precipitating stearic acid. Brisk stirring at low temperatures while the acid is in the process of crystallizing out may prevent the formation of relatively large crystals; this may make the cream thin and prevent the development of a pearly sheen.

Words and Expressions

mayonnaise [ˈmeɪəneɪz] *n.* 蛋黄酱

globules [ˈglɒbjuːlz] *n.* 小滴, 小球体

lipophilic [ˌlɪpəˈfɪlɪk] *adj.* 亲脂的, 亲油性

agitation [ˌædʒɪˈteɪʃn] *n.* (液体的)搅动, 摇动

moieties [ˈmɔɪətiz] *n.* 一部分

stearate *n.* 硬脂酸

opaque [oʊˈpeɪk] *adj.* 不透明的, 不透光的

Brisk [brɪsk] *adj.* 轻快的, 活泼的

Unit 3   Pigments and Dyes

After completing this unit, you should be able to:
1. Describe the difference between pigments and dyes.
2. Tell the classification of dyes.

Lesson 1   Pigments and Dyes (A)

1. Dyes

Dyes diffuse in solution, most often aqueous, into the substrate to be dyed and must be fixed there, either by chemical reaction with the substrate or by formation of insoluble salts and pigments.

Direct or substantive dyes are able to dye native or regenerated cellulose directly from neutral, aqueous solution, without previous treatment with a mordant. They are usually azo or sometimes anthraquinone dyes containing sulfonic acid groups. However, their fastness to liquids (water, sweat, laundering) becomes so poor as the depth of color increases that the dyes have to be converted into insoluble compounds by a subsequent treatment with cationic substances or complex formation with copper.

Acid dyes likewise possess one or more sulfonic acid groups which form salts with basic groups. They are used for the dyeing and printing of wool and polyamides, and also of silk, leather, base-modified polyacrylonitrile, and for foods in the form of the soluble sodium salts.

Dispersion dyes, which are relatively poorly soluble in water, are used to dye semi-synthetic or synthetic fibers such as polyester, polyamide, and polyacrylonitrile. The dyestuff is dispersed in the liquor but, as shown by the dispersion coefficient, has a greater affinity for the hydrophobic fibers than for water and is almost entirely absorbed by the fibers. Since it diffuses only into the amorphous parts of the fibers, the dyeing should be carried out at a temperature above the polymer's glass transition temperature.

Cationic or basic dyes contain positively charged nitrogen atoms. They are employed as water-soluble salts in the dyeing of polyacrylonitrile and acid-modified polyester fibers, and for paper and leather.

Vat dyes are pigments that are absorbed by the substrate in reduced, water-soluble form (leuco form, for example reduced with sodium dithionite) and there converted back into the insoluble colored form(the pigment) by oxidation, for example by atmospheric oxygen. Leueo vat dyes are also used as esters, which must be saponified after absorption and before oxidation.

Like vat dyes, developing dyes are only transformed into their final colored form once they have been absorbed by the substrate. Amongst the members of this class are naphthol AS azo dyes and phthalocyanine developing dyes.

Certain suitable azo dyes too react within the substrate. They are convened into metal complex dyes by chromium, cobalt, or copper ions. These complexes can also be used as pigments. However, their significance is declining for environmental reasons.

Reactive dyes contain reactive groups that form covalent bonds with corresponding groups in the substrate. Suitable residues include the hydroxy groups of cellulose, the amino and thiol groups of wool and silk, and the amine and carbonyl groups of polyamides.

2. Dye Technology

In synthesis, dyes form as $1\% \sim 10\%$ suspensions and are turned into an easily filtered form by salting out, pH adjustment, or temperature change. A number of intermediate steps are necessary to convert the dye into a commercial product such as a powder, granules, an aqueous fluid paste, or a concentrated solution.

After filtration, the dye has the form of a briquette with a solids content of $15\% \sim 60\%$. The addition of hydrotropic substances such as urea or glycols permits direct conversion to highly concentrated solutions. The dry dyes, with a residual moisture content of $0.5\% \sim 5\%$, are usually ground down to grain sizes of $1 \sim 50$ μm in impact mills before further processing.

Since individual batches may differ slightly in tint and shade owing to different by-products and salt content, other dyes must be mixed in to adjust the tint (toning) and salts must be added to correct the depth of color (cutting).

This addition of standardizing agents has the further advantage of improving the properties relevant to the application of the dye. Depending on the form of the commercial product, different agents are used; common ones are:

- neutral inorganic salts: sodium sulfate, sodium chloride;

- alkaline inorganic salts: sodium carbonate, sodium bicarbonate, trisodium phosphate;

- acidic standardizing agents: sodium bisulfate, amidosulfonic acid, oxalic acid;

- buffers: mono- and disodium phosphate;

- complex formers: ethylenediaminetetraacetic acid, polyphosphates such as sodium hexametaphosphate;

- nonelectrolytes: dextrin, sugar, urea, benzamide;

- antidustmg agents: mineral oil (combined with emulsifiers), phthalate esters, triacetin;

- dispersants, antifoaming agents, wetting agents, antigelling agents.

When dyeing with dispersion dyes, the rate of dissolution depends on the particle size and the crystal modification. Therefore, it is not enough to set the particle size of such dyes by grinding ($0.5 \sim 2$ μm); the dyestuff must be converted into the most stable crystal modification by forming by addition of emulsifiers and solubilizers during grinding, or by heating the aqueous suspension. If this is not done, even small deviations in the process parameters could lead to faulty dyeing.

Lignin sulfonates or the condensation products of naphthalenesulfonic acid and formaldehyde are used as dispersants, long-chain alkyl sulfonates or alkyl naphthalene- sulfonates are added to act as wetting agents, and various polyethoxylated products as emulsifiers.

The formulation of aqueous suspensions demands the addition of preservatives and of humectants to retard drying out (glycols, glycerin). If dispersion dyes are made commercially available in dried form, the very fine dispersion must be stable; a large excess of dispersant is therefore added to such products to prevent aggregation, in proportions of 1 : 1 or, better yet, 2 : 1 to the dye.

Concentrated solutions are gaining more and morepopularity as commercial forms, not only because the dosing is easier but also because the problems caused by dust can thus be avoided; inhalation of dye dust can result in allergies and illness for sensitive people. The solvent is chosen according to the intended substrate; the choice must take guidelines for environmental protection into consideration. Polyols (e. g. ethylene glycol), ether alcohols (e. g. diethylene glycol monoethyl ether), and, depending on the class of dye, carboxylic acids (e. g. acetic acid), acid amides (dimethylformamide, urea) are used as solvents or solubilizers for addition to water.

Words and Expressions

pigments [ˈpɪɡmənts]　n. 色素, 颜料

saponify [səˈpɒnəˌfaɪ]　v. 使皂化

dyes [daɪz]　n. 染料, 染液

phthalocyanine [ˌθæləˈsaɪəˌnɪn]　n. 酞菁染料

regenerated [rɪˈdʒenəreɪtɪd]　v. 再生的

cellulose [ˈseljuloʊz]　n. 纤维素

mordant [ˈmɔːrdnt]　n. 媒染, 媒染剂

granule [ˈɡrænjul]　n. 小粒, 颗粒, 细粒

azo　n. 偶氮基, 偶氮

briquette [brɪˈket]　n. 坯块, 压坯, 模制试块

anthraquinone　n. 蒽醌, 烟华石

residual [rɪˈzɪdjuəl]　残数的, 残余的

fastness [ˈfæstnəs]　n. 牢度, 坚牢度

moisture　潮湿, 湿气, 水分

amorphous [əˈmɔːrfəs]　adj. 无定形的, 非结晶的

tint [tɪnt]　n. 色调, 浅色, 色泽; v. 给……染色

fiber [ˈfaɪbər]　n. 纤维, 纤维质

triacetin [traɪˈæsɪtən]　n. 甘油醋酸酯, 乙酸甘油酯

dithionite　n. 连二亚硫酸盐

formaldehyde [fɔrˈmældəˌhaɪd] *n.* 甲醛, 福尔马林

substantive dyes 直接染料

dispersion coefficient 分散系数, 分散率

depth of color 色泽深度, 色浓度

dispersion dyes 分散染料

glass transition temperature *n.* 玻璃态转化温度

leuco vat dyes 隐色还原染料

positively charged *adj.* 带正电的

residual moisture 残留水分

polyester fibers *n.* 聚酯纤维

take into consideration 考虑到

Vat dyes 还原染料

Exercises

1. Give a few of examples of the dyes and pigments in our daily life and list the main components in those dyes and pigments.

2. Describe the differences between the dyes and the pigments.

3. Search the English literatures about dyes and pigments and describe the development trend of dyes and pigments.

Reading Material: Pigments and Dyes (B)

1. Solubility of Pigments and Dyes

The difference between pigments and dyes lies in their solubility. Pigments are practically insoluble in the medium for application (the vehicle) and are dispersed in it as solid particles, usually of size less than 1 μm. Dyes, in contrast, are absorbed by the substrate, for instance a yarn, in dissolved form and then fixed there by hydrogen bonding or chemical reaction with the substrate and the formation of poorly soluble salts or pigments so that it is difficult or impossible for the dye to diffuse out of the substrate.

Dispersion colors are a borderline case; they are used in aqueous dispersions at high temperatures, say 130 ℃ under pressure. At this temperature they are sufficiently soluble in water to permit diffusion in the dyeing process (Table 1).

Table 1　Solubility in water of dispersion colors and pigments

	Solubility in water at 130 ℃/(mg/L)
Dispersion colors	5 ~ 500
Pigments	< 0.05

Sometimes pigments are classified as either classical or high-grade pigments, high-grade pigments are often superior to classical pigments in some respects concerning their application, such as their stability to migration and recrystallization, thermal stability, lightfastness and fastness to exposure (to weathering). This is principally a result of their very low solubility.

Whether classical or high-grade pigments are used depends upon the application. Classical pigments with their higher solubility cannot be used in solvent-containing baked enamels (though they are suitable for use in air-drying alkyd resin coatings). At that high temperature, crystal growth, bleeding, and blooming can occur.

2. Characterization of Pigments

The application characteristics of pigments are largely determined by parameters related to the arrangement of the molecules in the pigment crystal, crystal shape and crystallinity, specific surface area, the nature of the surface, and the chemical properties of the surface. The morphology of the powder, aggregate, and agglomerate forms are also significant. These parameters can be changed, conditioned, in such a way as to optimize the technical characteristics for specific applications.

Synthetic products can bereduced to the desired particle size by grinding, and the crystallinity thus disrupted can be restored by gentle recrystallization or thermal treatment. Sometimes the surface of the pigment must be chemically modified. this treatment can improve the pigment dispersibility and the interactions between pigment and binder.

3. Applications for Pigments

The term"paint" covers liquid or powdered preparations that are thinly coated onto surfaces and turned into permanently adhering coatings by drying, baking, crosslinking, or polymerization. Careful selection of the paint to suit the surface to be coated is just as important as the optimal preparation of that surface.

There are numerous coating procedures, such as brushing, spraying, two-component spraying, dipping, electrostatic spraying or dipping, drum coating, centrifugal coating, pouring, rolling, and powder coating. The technique of coil coating, in which rolled steel or aluminum sheeting is coated at rates of up to 150 m/rain in widths of up to about 2 m, has become economically important.

The huge range of paints available can be classified according to end use, e. g. automobile paints, wood varnishes, lacquers for the interior of food cans, according to application method, e. g. spray paints, dip coatings, according to their form, e. g. solvent-containing, water-thinnable, powdered, or according to their drying characteristics, e. g. baked, air-dried.

Paints contain numerous components with defined tasks in the liquid for application and in the finished coating: volatile solvents and nonvolatile components such as binders (film-formers, resins, plasticizers) , additives, dyes, pigments, fillers.

Macromolecular substances such as nitrocellulose or vinyl chloride vinyl acetate copolymers are employed as film formers, or else substances of low molecular mass which polymerize as the paint dries are used, for example unsaturated polyester resins or epoxy resins. The viscosity of a polymer solution increases with the molecular mass of the polymer, so in industry low-viscosity coatings, in which the film-forming polymers only develop as the coating hardens, are preferred. These low molecular mass components are often liquids, so such coatings require little or even no solvent.

Nevertheless, even these coatings do require some high molecular mass components. Dispersants and dispersion stabilizers ensure the deagglomeration of powdery additives like pigments and fillers during the dispersion process in manufacture, and impede their flocculation in the prepared paint and on application.

The term"resins" is used for a group of film-forming substances of resinous consistency, which as a rule are readily soluble and are used to increase the solid content of paints, as well as enhancing their adhesiveness and theft gloss. In addition, resins increase the hardness of the film and shorten the drying time for systems that cross-link by oxidation.

Plasticizers are nonvolatile organic liquids of oily consistency, such as dioctyl phthalate. Assorted auxiliaries like driers antiskinning agents, hardening accelerators, running improvers, sedimentation inhibitors, matting agents, wetting agents, and antiflooding agents are added to improve the properties of the liquid paint or the finished coating.

Unit 4   Cosmetics

After completing this unit, you should be able to:
1. Describe the definition and classification of cosmetics.
2. Tell the difference between the cosmetics and drugs.
3. List the application of cosmetics.

Lesson 1   Introduction of Cosmetics

Cosmetics are important to both men and women. They help us present an image not only to others but to ourselves. In a study of hair coloring and makeup, in which photographs were attached to fictitious work records for possible employment, it was found that hair coloring and makeup had a beneficial effect on offered salaries. The amount averaged 12% greater when the photograph showed the prospective employee after receiving the Cosmetic treatment rather than before. Most of us have more self-confidence facing others when we think we look our best. Cosmetics can cover skin blemishes and graying hair and can emphasize our best features.

But what is a cosmetic? What's in it? How safe is it? Is it worth its price?

The Food, Drug and Cosmetic Act defines a cosmetic as: "① Articles intended to be rubbed, poured, sprinkled or sprayed on, introduced into, or otherwise applied to human body or any part thereof for cleansing, beautifying, promoting attractiveness, or altering the appearance and ② Articles intended for use as a component of any such articles, except that such terms shall not include soap. "

To the United States Government, a cosmetic improves appearance, whereas a drug diagnoses, relieves, or cures a disease. However, things are not quite that simple. A product may be both a cosmetic and a drug if it is intended to cleanse, beautify, or promote attractiveness by affecting the body or if a cosmetic is also intended to treat or prevent disease. Examples of products that, by law, are drugs as well as cosmetics are dandruff shampoos, antiperspirants that are also deodorants and suntanning preparations intended to protect against sunburn. This has caused consternation among regulators sad the regulated because, for example, a sunscreen is considered a drug because it may protect against skin cancer and premature aging, whereas a tanning preparation is considered a cosmetic.

Another gray area is hormone-containing lotions. If the product affects the structure or function of the skin, then it is a medication. If the hormones are present in such small amounts that they have no effect, then it is sold under false pretenses because of the lack of effectiveness and the product should be considered misbranded.

Cosmetics have traditionally received little attention because it has been wrongly assumed that such products do not really affect our health and safety. The skin was believed to be a nearly perfect barrier that prevented chemicals applied to it from penetrating into the body. This belief went unchallenged until the 1960s when the much-heralded but unmarketed miracle drug DMSO proved its ability to carry substances with it through the skin and into the body's tissue and bloodstream. Until it was shown that rabbit's eyes were adversely affected by the drug, DMSO was being promoted as a through-the-skin carrier of all sorts of medication. In fact an increasingly popular new way to deliver drugs today is transdermally. Medication to prevent seasickness or to treat angina (chest pains) are placed in an adhesive dise for delivery through the skin.

It has now been accepted that all chemicals penetrate the skin to some extent, and many do so in significant amounts. There are good tests for skin penetration available, yet they are rarely used for cosmetics. Tests to determine the systemic effects or metabolic degradation of an ingredient are rarely performed either. And the most frequent cosmetic culprits were fragrances, preservatives, anolin and its derivatives, p-phenylenediamine, and propylene glycol.

The most difficult issue in the identification of ingredients and contaminants in cosmetics concerns cancer-causing agent. A number of cosmetic ingredients now in use have been shown to be carcinogenic in animals or to cause mutations in the Ames Test, which uses bacteria to determine genetic breaks in cells. Testing for mutagenicity—breaks in the genetic material of bacteria—can screen many chemicals inexpensively and rapidly. Almost all chemicals known to he carcinogenic in humans have been shown to be mutagens in these systems. However, not all mutagens are necessarily carcinogenic, Thus animal studies which cost approximately $ 300 000 to thoroughly test an ingredient are needed.

All agents that cause cancer in humans cause cancer in rats and truce, with the exception of trivalent arsenic. Whether all chemicals that cause cancer in animals also cause cancer in humans is not known, but many scientists in the cancer field believe they do.

Two contaminants found in cosmetic products have been shown to cause cancer. One is N-nitrnsodiethanolamine (NDELA), which penetrates easily through the skin when in a fatty base. It is a contaminant that appears to be produced by the interaction of two otherwise safe ingredients—for example, the amines (surfactants, emulsifiers, and detergents) and nitrites or nitroso compounds, such as in the preservative 2-bromo-2-nitro-propane-1, 3-diol (BNPD). The other carcinogen is 1, 4-dioxane, a contaminant of raw materials. It is believed that about one-third of the emulsion-based cosmetics containing polyoxyethylene derivatives have it in amounts ranging from 1% to over 25%.

After a brief introduction of cosmetics, we give shampoo as an example of cosmetic. Nowadays shampoos constitute one of the main products for personal care used by all strata of the population (age, sex…). In 1977 shampoos represented 43% of the total US haircare market.

Defined as "suitable detergents for the washing of hair, packaged in a form convenient for use", they have since undergone drastic changes in design and technology in order to respond to multiple requirements, from the most sophisticated to the simplest and which go much beyond the single purpose of cleaning.

As early as 1955, a panel reported that "women want a shampoo to clean and also to rinse out easily, impart gloss to the hair and leave it manageable and non-drying. " At the same time, it was also stressed that the principal trend evident in shampoo formulation is toward surfactant which have a milder effect on the skin. complete elimination of any sting if the product gets into the eye is the objective.

Conventional detergents of the anionic type seem to cause unpleasant after-effects to the hair roughly in proportion to their grease removing power, but many other materials which remove grease will cause no apparent deterioration in the hair condition. Thus, if the hair is extracted with ether or trichloroethylene it can rapidly be made essentially grease-free, but its condition will he found almost unchanged. The hair will still be smooth, lustrous, and easy to comb and set. Even among detergents, there will be found some which remove relatively little sebum, but leave the hair in bad condition, and others which wash quite thoroughly without harming the hair. So no cause-effect relationship is likely to exist between residual oil and hair condition and it is up to the cosmetic chemist to find the right balance between adequate soil removal and desirable hair condition.

This balance between cleaning and conditioning must be chosen with care, and it must also take into account the type of market to be supplied. People with greasy hair are highly critical of a shampoo which produces effects lasting only for three or four days, while those with dry hair may be more easily satisfied. However, many dry-haired people also use dressings of an oily character which must be removed, and in any case there is a deep-seated feeling, among women in particular, that the process of shampooing is cleansing, purifying activity, designed to free them from daily accumulation and maturation of grease, dirt, perspiration, cooking smells, dandruff, environmental pollution and so on. In fact, it seems sensible to retain the definition of a shampoo as a suitable detergent for washing the hair with the corollary that it should also leave the hair easy to manage and confer on it a healthy look.

These are the two basic attributes of the types of shampoo described here and it is these requirements, known as "conditioning effects" and "mildness", that shampoo formulators nowadays provide as the indispensable cosmetic counterpart to the original "claning" function.

However, no matter how grand a shampoo may be, its primary function remains that of cleansing the hair of accumulated sebum, scalp debris and residues of hair-grooming preparations. Although any efficient detergent can do this job, cleansing should he selective and should preserve a quantity of the natural oil that coats the hair and, above all, the scalp. Undesirable side-effects have been shown to occur when using some of the best cleansers and indeed some authors include cleansing among the functions of shampoo only as an afterthought, The view that shampoos should be "inefficient" detergents arises mainly from the theory that the after-effects of shampooing-difficulty in combing the hair, roughness to the hand, lack of lustre and "fly" when the dry hair is combed-are due to excessive removal of oil from the hair. This assumption is at first sight quite reasonable but further examination shows that it is very much oversimplified: if sebum somehow fulfills a natural function of protection, and enhances the lustre and lubricity of the hair, it also possesses the dangerous drawback of attracting and trapping dust and dirt and has a potentially deleterious effect on the maintenance of set and the "feel" of hair.

Words and Expressions

angina [æn'dʒaɪnə] *n.* 心绞痛

emulsifier [ɪ'mʌlsɪˌfaɪər] *n.* 乳化剂,乳化器

antiperspirant [ˌænti'pɜrsp(ə)rənt] *n.* 防汗剂

emulsify [ɪ'mʌlsɪˌfaɪ] *v.* 使乳化

arsenical [ɑr'sɛnɪkl] *n.* 砷化物; *adj.* 砷的,含砷的

fictitious [fɪk'tɪʃəs] *adj.* 虚构的,杜撰的,假造的

arsenide [ɑː'senaɪd] *n.* 砷化物

lanolin(lanoline) ['lænəlɪn] *n.* 羊毛脂

blemish ['blemɪʃ] *v.* 玷污; *n.* 污点瑕疵

mutagen ['mjutədʒən] *n.* 诱变

carcinogen ['kɑrs(ə)nəˌdʒen] *n.* 致癌物

mutagenic [ˌmjutə'dʒɛnɪk] *adj.* 诱变的

consternation [ˌkɑnstər'neɪʃ(ə)n] *n.* 惊愕,惊恐

perspiration [ˌpɜrspə'reɪʃ(ə)n] *n.* 排汗,出汗

contaminant [kən'tæmɪnənt] *n.* 沾染物

p-phenylenediamine 亚苯基

contaminate [kən'tæmɪˌneɪt] *v.* 弄脏,沾染,毒害

premature [ˌpriːmə'tʃʊr] *adj.* 早熟的,不成熟的

cosmetic [kɑz'metɪk] *n.* 化妆品; *adj.* 化妆用的,整容的

preservative [prɪ'zɜrvətɪv] *n.* 防腐剂,防腐料

cosmetology [ˌkɑzmə'tɒlədʒɪ] *n.* 整容术

rinse [rɪns] *v.* 冲洗,清洗(发,手),漂净

dandruff ['dændrəf] *n.* 头垢,头皮屑

seasick ['siˌsɪk] *adj.* 晕海,晕船

dermal ['dɜːməl] *adj.* 真皮的,皮肤的,表皮的

seasickness ['siˌsɪknɪ] *n.* 晕海,晕船

detergent [dɪ'tɜrdʒənt] *n.* 清洁剂,去垢剂

sebum ['sibəm] *n.* 皮脂,脂肪

diagnose ['daɪəgˌnoʊz] *v.* 诊断(疾病),分析

sprinkle ['sprɪŋk(ə)l]   *v.* 洒, 喷淋, 撒布; 洒, 喷淋

1,4-dioxane   1,4-二氧杂环己烷, 二噁烷

transdermally [træns'dərməli]   *adv.* 经过真皮的

trichloroethylene [ˌtrai.klɔːrəu'eθiliːn]   *n.* 三氯乙烯

much-heralded   事先大肆宣传的, 广告先行的

attach to   把……放在

deep-seated   根深蒂固的, 深层的

are present in   存在于, 出现于

Exercises

1. Give a few of examples of the cosmetics in our daily life and list the main components in those cosmetics.

2. Describe the differences between the cosmetics and the drugs.

3. Search the English literatures about cosmetics and describe the development trend of cosmetics.

Reading Material: The Application of Cosmetics

1. Introduction

The correct use of cosmetics falls into two parts:

(1) Skin care, which has the maintenance of a soft, supple and clean skin and the prevention of effects due to external causes such as excessive exposure to cold, heat, sun, wind, etc.

(2) Decoration to produce a pleasing appearance by minimizing facial defects of colour or shape and directing attention towards better points.

The first depends to a large extent upon the type and condition of the skin. The possessor of a healthy normal skin is fortunate in that it will withstand many treatments and conditions which can have serious effect on skins which are definitely dry or greasy and which demand particular care and treatment. Even the healthy and normal skin may vary from time to time and need particular care to correct any departure from normality.

The second point—the amount of cosmetic decoration of make-up tolerable—is dependent chiefly on social conventions which vary from time to time and from society to society. The general trend for many years has been towards a greater tolerance, although there have been shorter swings in particular social sets to exaggerate effects in various directions; these are generally short-lived, but are assimilated, often in severely modified form, and contribute to the general trend.

2. Care and Cleansing of the Skin

It goes without saying that a healthy skin should be a clean skin, and in general a good toilet soap and water is the best way of achieving this cleanliness. A final rinse with clean cold water is an excellent measure to promote circulation and tone up the skin. At night such cleansing should be followed by the application of a cream of more or less greasy type according to the natural dryness or greasiness of the skin.

The cream should be applied by proper massaging movements, whether or not the massage is carried on for long, because friction increases the supply of blood to the skin and skillful massage assists in keeping the skin supple. Charts designed to show the paths which should be followed in facial massage and application of cosmetics have been prepared by cosmetic manufacturers (see Figure 1).

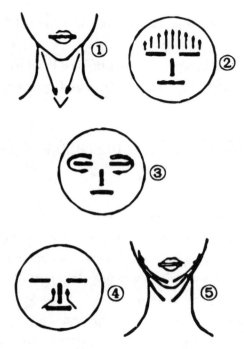

Figure 1　Facial Massage

The regular use of a correctly formulated face cream in suitable manner can prevent the premature (not ultimate) aging of the skin due to external causes and delay the appearance of wrinkles produced by loss of epidermal elasticity and subcutaneous moisture. It cannot, however, prevent the natural aging of the skin due to metabolic processes nor wrinkling caused by illhealthy, psychogenic factors or certain disease.

Wrinkles must be differentiated from the lines produced by emotions, for example constantscowling or sneering. The best cure for these is to cultivate a placid disposition.

In the case of excessively dry or greasy skin it may be necessary or advisable to cleanse it without the use of soap. The appropriate choice of material for dry skins would be complexion milk or a more greasy cleansing creams; for greasy skins, non-greasy cleansing lotions based on very mild alkalis such as sodium bicarbonate, or on certain buffered soapless detergents, may be used with advantage. Whatever method is adopted, the cleansing should be efficient and no trace of the day's grime or make-up should remain.

(1) *For the neck*—Using alternately both hands cupping round the neck, firmly massage with downward movements.

(2) *For the forehead*—Gently massage the forehead upwards from the eyebrows to the hairline with alternate movements of the hands.

(3) For the eyes—Both hands working simultaneously, make circles round the eyes, ending by pats on the lower eyelid. Always work in the direction indicated on the diagram.

(4) *To smooth the "smile line"*—Massage with both hands with upward symmetrical movements over the "smile line", finishing between the corner of the mouth and nostrils.

(5) *For the lower part of the face*—Both hands working alternately, taking the jaw line between the second and third finger, make upward movements from the point of chin to temporal muscle.

3. Foundation

When applying make-up, the foundation, with a degree of greasiness appropriate to the condition of the skin, is applied first of all, just sufficient to provide an adherent base for the rouge or blusher and powder. It is applied in the same directions as shown in Figure 1 for massage, but with more gentle smoothing movements, extending well on to the throat and further if desired, to avoid any lines of demarcation. It also hides skin imperfections and imparts a smooth, even appearance to the skin. In the case of normal or dry skin the foundation is usually a cream, but liquid foundation and cake make-up can equally be employed. In the case of oily skin, non-oily liquids, cake make-up and medicated lotions are more appropriate.

4. Rouge or Blusher

The application of rouge comes next. The rouge chosen should match the colour of the natural flush which appears through blushing or exertion. If it is applied correctly, it can accentuate the more attractive features and reduce the less attractive ones. Liquid or cream rouge is applied in sufficient quantity to the most prominent part of the cheek bones, just below the eyes and is spread with the finger tips, roughly into a triangle shape. The exact proportions and position of the triangles depend upon the shape of the face and the impression of length and breadth it is desired to convey. Naturally the edges of the coloured patch should blend easily into the remainder of the face with no hard outline. Powder rouge, however, is applied over the face powder.

5. Powder

The powder should be applied generously with cotton wool or a clean puff and the excess brushed off with a soft brush. It should be slightly darker than the foundation andhave a covering powder appropriate to the type of skin—light for dry skin, and heavy for an oily skin.

Thought should be given, as with all cosmetics, to the lighting under which the make-up will be seen; daylight demands shades nearer to natural, while for artificial light all colours should be shade. To obtain a very transparent effect, which can look attractive especially for evening wear, two contrasting powders should be used, first a lighter one and then a deeper one. With two shades of powder used simultaneously, the same effect can be achieved as by the location of rouge, namely a change in the apparent shape of the face. Applying a powder of a darker-shade across the lower part of the face will appear to shorten a long face. By application of a lighter-shade powder to either side of the jaw line, a heart-shaped face will appear wider.

6. Eye Make-up

For a long period eye make-up was practised much more discreetly by European of American women than by their Eastern sisters; even after the emancipation of make-up, lips were made up far more prominently than eyes. However, the pendulum has now swung and all aspects of this most attractive feature of a woman's face are enhanced by the use of the appropriate cosmetics, outdoors by day as well as indoors by night. Thus, there are pencils to improve and enhance the colour and shape of eyebrows, mascara to colour and lengthen the lashes, eyeshadow to draw attention to and re-shape the eyelids and to enhance the eye colour and eyeliners to outline and emphasize the eye itself.

These are applied as follows.

Pencil. The eyebrows are first of all tidied, if necessary, by slightly plucking them with tweezers, a thin line should, however, be avoided. The eyebrows are then prolonged deftly towards the temples by means of an eyebrow pencil which should match the natural colour of the eyebrows. In order to obtain a neat result, a sharpened pencil should be used. The pencil can also be used to outline the upper eyelids just above the eyelashes, again extending the line slightly towards the temple.

Mascara. The function of mascara is to increase the natural charm of the features by darkening the eyelashes and increasing their apparent length. It is claimed that, by its judicious use, the brightness and expressiveness of the eyes is enhanced.

Mascara is available in either cake or cream form, and slightly different techniques are employed in their application. Cake mascara is applied by rubbing the wetted brush over the cake until sufficient colour has been imparted to it and until it is almost dry. It is then stroked over the upper lashes only, whereby the colour is concentrated on the tips and outer lashes in order to secure a natural effect. A fresh brush is used when the lashes are dry in order to separate them.

At one time cream mascara had to be applied by means of the finger, and consequently was much less popular than blocks. However, the advent of a much more convenient applicator, in which the mascara as a thin cream is packed in a cylindrical container and applied by means of a small-diameter cylindrical brush, has increased the popularity of cream mascara to such an extent that it is now the dominant type. The brush, charged with mascara, is revolved against the outer side of the eyelashes, working away from the eyelid. The brush simultaneously deposits mascara and separates the eyelashes. Black mascara is suitable for black or dark brown eyelashes, while blue mascara when applied to the lashes of blue or grey eyes increases their apparent blueness.

Eyeshadow. Eyeshadow is used to impart more depth to the eyes and to intensify their colour. It is available in cream and stick forms, or as a pressed powder. The shade is governed by the colour of the eyes, and may include colours ranging from black and blue to green and silver. The cream eyeshadow is applied just above the lashes on the centre of the eyelids and smoothed in with an outward movement of the finger tips. Pressed powder eyeshadows are usually sold with a sponge-tipped applicator and this is used to transfer the product from its compact onto the eyelids. Some powdered eyeshadows can also be applied with the finger. For daytime wear, eyeshadow should be applied lightly, and for evening wear the application should be only slightly more heavy. The use of two products at the same time is becoming very fashionable. A dark colour (blue, green or brown) is applied on the eyelid and lighter coloured product or "high-lighter" (pearlized white or pink) is applied in the gap between the eyelid and eyebrow to merge and tone in with the original colour. Wearers of heavy-framed spectacles should refrain from heavy eye make-up.

Eyeliner. According to the instructions issued with one of the many liquid eyeliner preparations, the brush provided with it is dipped into the liquid, held as one would a pencil, and then the desired line is drawn in one single brush-stroke. The same instructions also state that an eye opening effect can be achieved by broadening the line towards the centre of the eye.

7. Lipstick

The last touch to the make-up, but by no means the least, is the lipstick, which is used to impart colour and an attractive shape and appearance to the lips, and also to protect them. Intelligently used, it is capable of altering the apparent facial characteristics. Thus, by using a coat of darker lipstick with a lighter shade on top, narrow lips can be widened; by other methods wide lips can be narrowed, and the length of the lips can be brought into better proportion to the shape of the face.

The desired mouth shape is first outlined and then filled in. The lips are blotted by biting on a tissue and the lipstick is applied once more. After a few seconds the lips are pressed together without tissue and the make-up is complete. If a more matt effect is desired, the lips should be lightly coated with powder before they are finally pressed together.

The colour of the lipstick can vary more than that of almost any other cosmetic and is chosen by the user not only on the basis of the tone of the skin, but also depending on the colours of the clothes worn. The intelligent manufacturer will keep abreast of such fashion changes and change his lipstick range accordingly.

Words and Expressions

matt [mæt] *adj.* 不光亮的,无光泽的,哑光的

judicious [dʒu'di,ʃəs] *adj.* 明断的,有见地的

subcutaneous [ˌsʌbkju'teɪniəs] *adj.* 皮下的

lashes [læʃz] *n.* 眼睫毛

mascara [mæ'skerə] *n.* 睫毛膏

cylindrical [sɪ'lɪndrɪk(ə)l] *adj.* 圆柱形的,圆筒状的

tweezers ['twizərz] *n.* 镊子,小夹钳

Reading Material: Classification of Cosmetics and Toiletries

Experts recently submitted for discussion the classification of cosmetic preparations shown in Table 1.

Table 1　Classification of cosmetics and toiletries

1. Skin care products		2. Cosmetic products with specific efficacy	
Bath products	Skin cleansers	Sunscreen preparations	Skin tanning preparations
Skin care products	Eye care products	Skin bleaches	Insect repellents
Lip care products	Nail care products	Insect bite lotions	Deodorants
Feminine hygienic products		Antiperspirants	Acne care products
3. Oral care products		4. Hair care products	
Oral hygiene products	Denture cleaners	Shampoos	Hair care products
Denture adhesives		Hair setting products	Hair waving products

Skin cleaning preparations. The oldest skin cleaning product is undoubtedly soap; it was already being used in the form of wood ash solution for the washing of the body and clothing approx. 4 500 years ago by the Sumerians and Egyptians, and later by the Germanic tribes and the Gauls. The belief, which unfortunately, is still prevalent today, that water and soap is adequate for the cleaning of the skin and hair, is no longer compatible with modern society's demand for personal hygiene and cleanliness. Water was described as the agent "that washes all harm away", but based on current knowledge it is no longer considered the ideal basis for a washing solution. Exposure of the skin and its appendages to water over a period of time will cause a swelling of the stratum corneum or the hair surface. This swelling results in the progressive elution of water-soluble substances present in the stratum corneum, which is very important for the moisture retention capacity, and finally to an increase of the permeability of the stratum corneum. Furthermore, water is not a suitable cleaning agent, because of its inability to adequately remove fats and fat-like substances.

Neither is soap an ideal cleaning agent. The washing solution formed by soap in presence of water is strongly alkaline (pH value 9 ~ 11), causing an increased swelling of the stratum corneum when compared to the effect of pure water. Soaps may thus lead to skin irritations after frequent or long contacts. Moreover, like other surface active substances, soap not only removes the protective skin surface lipid films, but simultaneously releases appreciable amounts of fatty substances and constituents responsible for the moisture retention capacity from the stratum corneum. The quality, and also the skin compatibility of the soaps intended for body care, can be substantially improved by the application of purified fatty acid fractions, selected for better skin compatibility. The skin compatibility of soaps can be improved further by the addition of superfatting agents or of special skin-protecting additives.

Today surfactant solutions (liquid soaps) are utilized as a soap replacement. They are aqueous, or aqueous/alcoholic solutions which contain one or several surfactants, and have been adjusted to a certain viscosity. Surfactant solutions that have been adjusted to higher viscosity are called washing gels.

The skin-cleaning preparations that provide a particularly mild cleaning action should also be mentioned. Skin compatible and wash effective components are utilized in their preparation. These preparations are frequently recommended for cleaning skin areas which are especially sensitive inflammation and which are irritated by regular soap.

Skin care preparations. Skin care preparations are products which are applied for general skin care and may be intended either for whole-body care or for the care of specific areas only. They should replace to a great extent the skin surface lipid film that has been removed by bathing (cleaning) and provide the skin, predominantly the stratum corneum, with substances that had been removed or eluted during cleaning (e. g. the moisturizers). These preparations should also protect the skin against the damaging influence of sun light, and other environmental influences.

A multitude of products exists that is capable of fulfilling the functions just described. Foremost are skin creams. These are W/O or O/W emulsions withvariable water content. They allow the application and uniform distribution of skin care effective fats and fat-like ingredients in emulsion form over large skin areas. The great variety of suitable fatty base materials for the preparation of skin emulsions makes it possible to produce compositions of the desired consistency (viscosity), which are able to hold incorporated active ingredients in a stable form and to permeate even the horny layer of the skin.

Hair care products. For the assessment of the appearance of a fellow human being, the condition of the beard hair, hair color, and also the fit of the hair style have always been an essential factor. For this reason, a special importance is attached in the cosmetic area to the hair care products. Up front are those products, that make the cleaning of the hair possible, and such that help to improve the styling effect.

Most shampoos contain a combination of surfactants which provide a mild cleaning of hair and scalp. A. L. L. Hunting evaluated the composition of 438 shampoos available in the USA and determined that a shampoo contains on average approx. 15 substances. Included are various surfactants, materials that prevent dandruff, oily, or brittle hair and additives that condition the hair during washing. Some of the substances included to achieve these results are protein hydrolysates quaternary ammonium compounds and their derivatives, as well as various other additives.

Hair colorants generally contain cationic substantive dyestuffs (direct dyes) such as the nitro dyes, or the precursors of oxidation dyes. The latter are mostly aromatic p-diamino or p-hydroxyamino compounds, the so-called developers, and m-diamino, m-amino hydroxy and m-dihydroxy compounds, the so-called coupling agents. These precursors react in the hair together with oxygen from hydrogen peroxide or solid hydrogen peroxide addition compounds to form oxidation dyes.

Hair tints for the covering of light to middle gray hair contain, in addition to fatty alcohols and their derivatives, mild surfactants such as fatty alcohol sulfates or fatty alcohol ether sulfates.

Hair coloring preparations intended for the dyeing of severely gray to white hair are of similar composition as the hair tints, but do not contain surfactants which decrease the effectiveness of the dyestuff.

Shaving preparations. These are products which are:

- applied before shaving to prepare the beard hair for the subsequent wet or electric shave, or

- applied after shaving to overcome the "damaging" effects of the shaving procedure, especially of wet shaving, on the stratum corneum.

Words and Expressions

matt [mæt] *adj.* 不光亮的, 无光泽的, 哑光的
judicious [dʒuˈdiʃəs] *adj.* 明断的, 有见地的
subcutaneous [ˌsʌbkjuˈteiniəs] *adj.* 皮下的
lashes [læʃz] *n.* 眼睫毛
mascara [mæˈskerə] *n.* 睫毛膏
cylindrical [siˈlindrik(ə)l] *adj.* 圆柱形的, 圆筒状的
tweezers [ˈtwizərz] *n.* 镊子, 小夹钳

Reading Material: A Two-for-One Deal—Changing the Look of the Hair Care Industry

The time-conscious, cost-conscious, environmental conscious 1990s are spawning a slew of new products in the household and personal care areas.

One of the newer shampoo categories to capitalize on these characteristics of the 1990s consumer are the shampoo/conditioner combinations, which are now housed in one product, in one bottle. The appeal of such a product is easy to see: less time spent on cleaning and conditioning the hair, less cost when buying one product instead of two and less waste then disposing the empty bottle.

Though the concept of combining conditioner with shampoo is not a brand-new idea, the explosion of products on the market is a more recent phenomenon. Most industry experts concede that the introduction in the early part of the 1980s of Pert from Procter & Gamble was the first major mass market attempt at capturing an audience that was tired of the two step cleaning conditioning process. Though the concept wasn't revolutionary, the wide-spread success of Pert was.

But even though the product proved successful, manufacturers still stayed with the tried-and-true two step process of cleansing andconditioning. They updated shampoos and conditioners, fine tuned them for hair types and cleansing requirements, and continued to sell the consumer on the idea of buying separate products to fit hair needs.

However, by the beginning of this decade, the consumer was changing dramatically with the times, and interest was becoming demand as consumers looked for products that fit not only their hair type and style, but their lifestyle and beliefs as well. According to industry analysts, the shampoo-plus-conditioner-in-one segment now accounts for approximately 8. 5% of the $ 1. 7 billion shampoo market.

Though for most manufacturers, their product isn't being marketed from an environmental standpoint, they are still fairly quick to point out that the convenience of putting shampoo and conditioner in one bottle is helping the country's landfill problem. Vidal Sassoon Ultra Care System from Richardson-vicks, Inc. , which incorporates Sassoon's three-step system of a salon shampoo, a protein conditioner and a protective finishing rinse, comments, "Shampoo, conditioner and finishing rinse in one bottle means less waste to clutter the environment. "

For the people at S. C Johnson & Son, Inc. , however, this ideal wasn't progressive enough. The importance today, notes a spokesperson for the company, is source reduction. "Plastic recycling is still in its infancy, " she comments.

One of the most important methods of source reduction is designing packages to contain as little excess material as possible.

S. C Johnson believes they have done just that with the launch of Agree Plus, its new shampoo/conditioner. The product, notes the spokesperson, was developed to be competitive in the marketplace with such products as Procter & Gamble's Pert Plus. "Agree is a very strong and successful name in the shampoo market segment, and the combination shampoo and conditioning item offers convenience and cost efficiency, which is what the customer is looking for, " she explains.

However, it is not so much the product as it is the packaging that is making the launch so interesting. "Agree Plus is the first and only liquid dispensing stand-up pouch to debut in the U. S. Patented refinements incorporated in the enviro pouch package—including the no-drip spout, and 'pinch points' near the top of the package-represent improvements over pouches in other countries that are now being used in refill capacities, " notes the company. The spokesperson also says it is the first time this type of pouch is available for a consumer product.

From a marketing standpoint, the environmentally-conscious consumer is also paring down other elements, and the two-in-one shampoo fits right into these lifestyle changes. Time and cost are big issues now for the consumer, and two-in-one shampoos address them both.

Ease of use is the focus of Procter & Gamble's Head & Shoulders two-in-one formula. According to the company, this formula was developed for "consumers who want the benefit of shampoo, complete conditioner and scalp relief in one easy step. "

All of P&G's two-in-one shampoo products, including Pert Plus, Head & Shoulders and Vidal Sassoon Ultra Care, have patented formulations.

"Because not all two-in-one products are suitable for all hair types, Rave All in One Shampoo and Conditioner offers three formulations—Normal to Oily, Dry and Damaged, Permed or Color-Treated—each with its own level of conditioning, " notes Roballey. All three formulations, which were launched early, last year, contain oil-free conditioner.

Head & Shoulders two-in-one comes in three formulas, including conditioner: Dry Scalp shampoo, which helps provide relief of dry scalp; Original Dandruff shampoo, which helps provide dandruff protection; and Intensive Treatment Dandruff shampoo, which helps provide persistent dandruff protection.

"The approach of Agree Plus is a little different from the competition, " notes S. C Johnson's spokesperson, "in that it offers three formulations—light, normal and extra conditioning—which concentrate more on conditioning rather than cleansing. " All three formulations are dye free.

Reading Material: Lipsticks

1. Introduction

Lipsticks, the lip cosmetics moulded into sticks, are essentially dispersion of colouring matter in a base consisting of a suitable blend of oils, fats and waxes.

Lipstick is used to impart an attractive colour and appearance to the lips, accentuating their good points and disguising any bad ones. Narrow bad-tempered lips may be widened, and broad sensual lips made to appear narrower by its use. In fact, if applied intelligently it is capable of entirely altering the apparent facial characteristics.

Since lips are considered to be more alluring when they possess a slightly moist appearance, this is always achieved by the use of a greasy base which also exerts an emollient action.

There is no doubt that the wide use of lipstick among women has led to a decrease in cracked and chapped lips, the crevices of which were always liable to bacterial infection. In addition, as in the case of many other cosmetics, it exerts a psychological effect difficult to assess, and induces a feeling of mental comfort.

2. Characteristics Required in a Lipstick

A good lipstick should have the following characteristics.

(1) It should have an attractive appearance, that is, a smooth surface of uniform colour, free from defects such as pinholes or grittiness due to colour or crystal aggregates. This should be retained during its shelf life and usage life—it should not exude oil, develop a bloom, flake, cake, harden, soften, crumble nor become brittle over the range of temperatures.

(2) It should be innocuous, both dermatologically and if ingested.

(3) It should be easy to apply, giving a film on the lips that is neither excessively greasy nor too dry, that is reasonably permanent but capable of deliberate removal, and which has a stable colour.

It will be realized that a system of colouring matter dispersed in a plastic fatty medium is one most likely to satisfy the above requirements.

3. Ingredients of Lipsticks

3.1 Colouring Materials

The colour of a lipstick is one of the major selling points, but it is one which can only be dealt with in general terms, since the precise shades are dictated by ephemeral fashion. It is usual for the colour to contain some measure of red and this allows shades ranging between orange-yellow and purple-blue, although even greens are not unknown. Depth of colour and opacity are also variable and during periods when the fashion trend was to a "no make-up" look, uncoloured lipstick base of high gloss has been seen under the name "lip gloss". "Lip gleams" containing pearlesent materials are also known, and occasionally sticks with some degree of gold or silver lustre achieved by the use of finely divided metal (coloured aluminium) are presented. However, the main accent in this chapter will be on the predominantly red conventional shades, since these involve the basic principles of lipstick formulation.

The colour is imparted to the lips in two ways: ① by staining the skin, which requires a dyestuff in solution, capableof penetrating the outer surface of the lips; ② by covering the lips with a coloured layer which serves to hide any surface roughness and give a smooth appearance. This second requirement is met by insoluble dyes and pigments which make the film more or less opaque.

Typical proportions for the colours in a lipstick are as follows: staining dyes (bromoacids) 0.5% ~3%, oil-soluble pigment 2%, insoluble pigment 8% ~10%, titanium dioxide 1% ~4%.

(1) *Staining Dyes*. The most widely used staining dyes are water-soluble eosin and other halogenated derivatives of fluorescein which are generally referred to collectively as "bromoacids", a term originally applied to acid eosin, tetra-bromofluorescein.

Eosin, also known as D&C Red No. 21, is an insolubleorange compound which changes to an intense red salt when the pH value is above 4. When applied to the lip- in the acid form, it produces a relatively indelible purple red stain on neutralization by the lip tissue.

Unfortunately, eosin and some of its derivatives can give rise to sensitization or photosensitization, leading to cheilitis or more general allergic reactions. Whether this is due to the bromoacid perse, or to impurities contained therein or to the perfume in the lipstick is by no means clear, but the fact that it does occur with a small proportion of lipstick users has focussed attention on permissible colours.

(2) *Pigments*. Both inorganic and organic pigments and metallic lakes are used to give intensity and variation of colour. When selecting lakes the possibility of reaction with the base, for example soap formation with free fatty acid, must be borne in mind.

①*Lakes* of many of the D&C colours with metals such as aluminium, barium, calcium and strontium are potential pigments for lipsticks. However, some strontium and zirconium lakes have to be avoided in most EEC countries, as they are banned. Aluminium lakes are not usually favoured because of their lack of opacity, but this very property would seem to suggest their use in transparent lipsticks.

The following lakes are considered to be the most useful lipstick colorants:

Calcium lakes of D&C Reds Nos. 7, 31 and 34

Barium lakes of D&C Red No. 9 and D&C Orange No. 17

Aluminium lakes of D&C Reds Nos. 2, 3, and 19 and FD&C Yellows Nos. 5 and 6

When the parent D&C colour is sbbject to restriction the lakes are also restricted in the same way.

Iridescent lipsticks utilize either mica coated with titanium dioxide or bismuth oxychloride at levels of up to 20%, depending upon the effect desired.

②Titanium dioxide, often used at levels up to 4%, is the most effective white pigment for obtaining pink shades and giving opacity to the film on the lips. However, the use of titanium dioxide requires great care in the grade of material selected (anatase or rutile) and the surface treatment it has received to make it lipophilic, and also in the method of incorporation, if unexpected troubles such as oily exudation, streaking, dullness and coarse texture are to be avoided.

3.2 *Base*

Apart from the colour, the quality of the lipstick during manufacture, storage and use will be determined for the most part by the composition of the fatty base. This quality is largely concerned with the rheology of the mixture at various temperatures. For instance, during manufacture (usually while warm) it must be possible to mill and grind the mass, and to pour and mould it while holding the insoluble colours evenly dispersed without settling. In the moulds it must set quickly with a good surface and good release properties. During shelf life and usage life the stick must remain rigid and stable, and generally in good condition. In use the stick must soften sufficiently in contact with the lips, and be sufficiently thixotropic to spread on the lips to form an adherent film which will not smear nor, transfer to cups or glasses.

(1) *Dyestuff Solvents*. Although all the base ingredients must contribute to the physical and rheological properties, there is the additional requirement that some part of the base must act as the necessary solvent for the staining dyes. Many of the normal fatty materials which might be considered for use in the base are too non-polar to dissolve the dyestuffs, and it is convenient to consider first those ingredients which do have solvent properties for eosin, and which must form some part of the base.

In general terms, vegetable oils have the greater solvent power for eosin but suffer from degradation properties. Mineral oils are more stable but have poorer solvent properties.

①*Castor oil* is a traditional material for dissolving bromoacid and it owes this property to its high content of ricinoleic (hydroxyoleic) acid, which is unique among natural oils. Its other properties include a high viscosity, even when warm, which delays pigment settling and the oiliness which helps with gloss and emollience, although too high a quantity causes drag during application and an unpleasant greasy film. As much as 50% has been used, but a better quantity is probably about 25%. The disadvantages of castor oil include an unpleasant taste and potential rancidity.

②*Some fatty alcohols* have some solvent power for dyestuffs. Any of these alcohols (lauryl-C_{12}, myristyl-C_{14}, stearyl-C_{18}, oleyl-C_{18} unsaturated, or cetyl-C_{16}), could be used according to the consistency contribution required. It is claimed that hexadecyl alcohol is a good solvent for bromoacid dyes, that lipsticks containing it can be applied with very little drag and do not bleed or smear and that its presence lessens any tendency to develop an unpleasant taste during storage.

③*Esters* of various kinds have been proposed, including lower alkyl esters of fatty acids and dibasic acids such as adipic and sebacic, short chain-length acid esters of fatty alcohols, mono-, di-, and mixed esters of glycol or glycerol. They have no specific virtues other than that they are lipophilic liquids of low oiliness giving lubricating and emollient effects with some dye solvent properties. If too much is used the stick may sweat.

④*Glycols* with two hydroxy groups are more polar than the fatty alcohols and might be expected to be better dye solvents. However, glycols are not particularly miscible with fatty materials and are of little importance.

⑤Polyethylene glycols (Carbowaxes) also have good dyestuff solvent power. The solvent power correlates to some extent with water solubility and this is a detraction. However, correctly chosen, these materials would seem to have considerable possibilities in lipstick formation.

(2) *Other Base Ingredients*. It will have been observed that few of the ingredients quoted so far have had the high melting points or hardness which are required to give satisfactory moulding properties, that is, quick setting and good release with a glossy surface and a rigid stick. This function is generally fulfilled by the inclusion of waxes or wax-like materials.

①*Carnauba wax* is a very hard vegetable wax used for raising the melting point, imparting rigidity and hardness and providing contraction properties in the moulding process.

②*Candelilla* is another hard vegetable wax serving the same functions as carnauba wax but has a lower melting point and is less brittle.

③*Amorphous hydrocarbon waxes* in mineral oil, for example ozokerite wax, give a short-fibred texture to the product.

④*Petroleum-based waxes*, for example microcrystalline wax, are also used to modify the rheology of the product.

⑤*Beeswax* is the traditional stiffening agent for castor oil, but it can give a grainy and dull effect if used in large quantities.

⑥ Cocoa butter might be thought an ideal material. However, it cannot be used alone since it does not have all the required properties, and used in too large quantities it will cause the stick to "bloom".

Other more or less wax-like materials are hydrogenated vegetable oils whichare more solid and less prone to rancidity than the unhardened oils. Of particular interest is hydrogenated castor oil, which in addition to wax-like properties still retains eosin solvent properties. Some other softer materials will be found to be of use in lipstick manufacture.

⑦*Lanolin and lanolin absorption bases* are very useful ingredients up to about 10% by virtue of their emollient properties. They are claimed to have eosin solvent properties and act as binding agents for the other ingredients, tending to minimize sweating and cracking of the stick and acting as plasticizers. Absorption bases in particular are recommended to enhance the gloss on the lips.

⑧*Petroleum jelly* and the more viscous *paraffin oils* may be used to adjust consistency, act lubricants and improve spreading properties. Large amounts tend to impair the adhesion properties and can be difficult to blend if much polar material such as castor oil is present.

⑨*Lecithin* is another possible component which acts as a dispersing agent for pigments, in addition to facilitating the application of the lipstick and improving the adhesion to the lips. From the foregoing description of the properties of the various materials it will be seen that no one or two materials are able to provide all the properties and qualities required in a lipstick; this serves to explain the almost invariable complexity of lipstick formulae, which is evident in the examples given below.

3.3 *Perfumes*

Special attention must be paid to the choice of perfume, which is frequently used in relatively high amounts (2% ~ 4%), from the point of view of consumer acceptance and freedom from irritation.

Perfumes selected should mask the fatty odour note of the base and should be nonirritant to the lips. Since the consumer is likely to apprehend the perfume in the mouth as well as the nose, the flavour must be considered as well as the odour. Perfumes should be stable and compatible with the other constituents of the lipstick base.

Component	Component Percent	
	(1) Percent	(2) Percent
Caster oil	30.0	50.0
Mineral oil	10.0	—
Beeswax	12.3	7.0
Paraffin	10.0	—
Carnauba wax	10.0	3.0
Ceresin wax	10.0	3.0
Silicone fluid (1 000cs)	10.0	—
Lanolin	—	10.0
Isopropyl myristate	—	5.0
Candelilla wax	—	7.0
p-Hydroxybenzoic acid propyl ester	0.2	0.2
Bromoacids	1.5	3.0
Colour lakes and pigments	6.0	11.8
Perfume	q. s.	q. s.

Words and Expressions

cheilitis   [ka͵ɪˈlaɪ͵tɪs]   *n.* 唇炎

ricinoleic   *n.* 蓖麻油酸

emollience   *n.* 软化作用

Iridescent   [͵ɪrɪˈdes(ə)nt]   *adj.* 色彩斑斓闪耀的

Carbowaxes   *n.* 碳蜡

thixotropic   [θɪksəˈtrɒpɪk]   *adj.* 触变性(的),触变性的

ozokerite   *n.* 地蜡

dyestuffs   [ˈdaɪ͵stʌfs]   *n.* 染料

carnauba   [kɑrˈnaʊbə]   *n.* 巴西蜡棕

eosin   [ˈɪəsɪn]   *n.* 曙红,类似曙红的染料

Ceresin   *n.* 地蜡

Unit 5　Flavors and Fragrances

After completing this unit, you should be able to:

1. List the catagories of flavors.

2. Describe the selection of fragrance.

Lesson 1　Citral

Citral is an extremely important aliphatic acyclic terpene aldehyde and, because of its unusual structure, it possesses properties which are of interest both from an academic and a practical viewpoint. The occurrence of citral in a wide variety of plants has led many investigators to believe that it plays a vital role in the biogenesis of essential oils. Citral has a pleasant lemon odor, and it is well known that the oil derived from lemon peel owes its characteristic odor largely to the presence of this aldehyde. Lemon oil contains only a small percentage of citral, however, and cannot therefore be considered a practical source of this compound. Fortunately, it is found to the extent of about 80% in lemongrass oil which is quite inexpensive, and most of the commercial citral is derived from this oil.

Although citral finds extensive application as a flavoring and perfumery ingredient, the major portion serves as a raw material for the manufacture of highly valuable compounds such as the ionones and methyl ionones. More recently, it has been used as the starting material for the synthesis of vitamin A and numerous other polyene compounds.

Citral is an alpha beta unsaturated aldehyde and, since it also has the typical terpenoid structure, it undergoes many interesting reactions. Because of its great sensitivity to acids and alkalies, experimentation with this aldehyde requires the greatest care and skill on the part of the organic chemist.

Physical Properties—When freshly distilled, citral is an almost colorless oil ossessing a characteristic lemon odor. The three isomers are claimed to have different odors, the gamma having the most pleasant and lemon-like character.

However, since traces of impurities influence the odor of a substance, and it is doubtful that any of these isomers have ever been prepared in absolute purity, the question of the relationship of odor to the constitution of the isomers needs further study.

Commercial Production—Citral is obtained from low-priced lemongrass oil which contains about 80% citral. The most common and most economical method of preparation involves the careful vacuum fractional distillation of the oil, and the fractions free from unpleasant odor impurities are marketed as citral. Such fractions usually contain 92% ~95% citral, the impurities consisting of aliphatic terpenes such as citronellal, citronellol, geraniol, etc. , the presence of which is not considered objectionable when the citral is used either for flavoring, perfumery, or for the manufacture of ionones. Citral prepared by this method should, however, be free of the characteristic, unpleasant note of lemongrass oil due mainly to the presence of methyl heptenone. Citral containing even traces of methyl heptenone lacks the fresh, lemon-like odor and cannot be used for either flavoring or perfumery purposes.

If a relatively pure product is desired, chemical methods of purification prior to rectification are employed. The method depends on the formation of a sulfite addition product which is hydrolyzed to yield the pure citral. On rectification of the liberated citral, a product is obtained which assays 98% or more aldehyde. Since treatment with sodium bisulfite tends to form stable sulfonates from which citral cannot be regenerated, it is general practice to carry out the process by using sodium sulfite solution, keeping the solution slightly alkaline. The modified methods of the procedure described by Hibbert are commonly used. This procedure is of considerable interest and is given below:

"A mixture of 1 750 g of crystalline sodium sulfite ($Na_2SO_3 \cdot 7H_2O$) , 625 g of sodium bicarbonate and 5 liters of water was agitated at room temperature for about two hours, or until the solid carbonate had disappeared. The solution was then cooled by means of an ice-bath to 5 ~ 10 ℃ and 500 g of commercial citral added with vigorous agitation, which was continued for three to four hours, intimate contact being effected between oil and solution, and air excluded as much as possible. Under these conditions, only the citral went into solution, namely, as the labile dihydrodisulfonic acid derivative. The solution was agitated thoroughly with about 500 ml. of ether, the ether removed and the process repeated in order to remove all undissolved impurities. "

In both the sulfite and bisulfite processes, losses of 10% or more occur, depending upon the care taken in carrying out the reaction. On the other hand, little loss occurs in preparing citral through a careful rectification under vacuum through an efficient fractionating column, and the process does not involve the use of special equipment and chemicals. Consequently, commercial citral is rarely purified by means of the sulfite process.

Words and Expressions

citral　['siˌtrəl]　n. 柠檬醛

terpenoid　['tɜːpənˌɔid]　n. 萜类,萜类化合物

terpene　['tɜːpiːn]　n. 萜烯,萜(烃)

impurities　[ˌimˈpjurətiz]　n. 杂质,不纯净物

biogenesis　[ˌbaiouˈdʒenˌsis]　n. 生源说,生源论

vacuum　['vækjuːm]　adj. 真空的,产生真空的

lemon grass　['lemənˌgræs]　n. 香茅草,柠檬草

citronellal [ˌsɪtrə'næl] *n.* 香芽醛

flavoring ['fleˌvərɪŋ] *n.* 调味品,香料

citronellol [sˌɪtrə'nelɒl] *n.* 香茅醇

ionones ['aˌɪənoʊnz] *n.* 芷香酮,紫罗(兰)酮

geraniol [dʒə'reˌnˌɪoʊl] *n.* 香叶醇

polyene ['pɒliˌin] *n.* 多烯,聚烯

phenolic note 酚气息

plays a role in 在……中起一定作用

pleasant note 愉快香韵

essential oils 香精油

rectification under vacuum 真空精馏

to the extent of 到……的程度

Exercises

1. Give a few of examples of the flavors and fragrances in our daily life and list the main components in those flavors and fragrances.

2. Describe the differences between the flavors and fragrances.

3. Search the English literatures about flavors and fragrances and describe the development trend of flavors and fragrances.

Reading Material: Coumarin

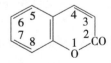

$C_9H_6O_2$ Mol. Weight 146. 14

Coumarin is one of the most important aromatic compounds, and its annual production runs into very high figures. It is one of the few chemicals used extensively not only in perfumery but to an even greater degree for flavoring purposes.

Because of its characteristic pleasant odor, it is used in compositions known as New Mown Hay and Fougere. Both odors are considered the woody type and are much used in men's toiletries and similar products.

The largest use of coumarin is probably in artificial vanilla. In many of these artificial vanilla compositions, tonka bean extracts, rich in coumarin, are employed. Large quantities of coumarin are used by the bakery industry to impart a pleasant and palatable vanilla-butter note to cakes, biscuits, etc.

The beverage industry is another user of coumarin, the flavor of at least one popular beverage (Cream Soda) being largely due to the presence of this compound and vanillin.

Many soap perfume formulas contain a small percentage of coumarin. The tobacco industry also utilizes coumarin to impart a pleasing aroma to smoking tobacco.

More recently, coumarin is being used in ever-increasing quantities as a neutralizer of disagreeable odors of many industrial products, such as rubber and plastic materials, as well as in household articles such as curtains, aprons, etc.

Limited quantities are also used in paints and sprays.

Occurrence—This compound was first discovered in 1820 in Tonka beans (Dipteryx odorata) and has subsequently been reported in a very large number of plants. A review made by Spath revealed the presence of coumarin in not less than 66 plants belonging to 24 families. Many derivatives of coumarin have also been found in various plants.

Some of the more important sources are listed below:

Melilotus officinalis D. and in others of melilot species.

Several species of orchids (Orchis militaris L.) and others.

Woodruff (Asperrula odorata) .

Oil of Lavender obtained by extraction process.

Oil of Cassia (Cinnamomum cassia P.)

Oil of Balsam Peru (Myroxylon balsamum, Pereirae) .

Coumarin usually Occurs in plants as a complex substance in combination with sugars or acids. It is liberated during the processing of the plant. Manufacture-For a long time after its discovery, the Perkin synthesis of coumarin served for the industrial production of coumarin and modified procedures are employed to this day. It was reported that the addition of very small amounts of iodine to the Perkin reaction mixture could appreciably shorted the reaction time. The reaction was still far from satisfactory, however, from the point of view of industrial manufacture. There always resulted considerable quantities of tar and resinous material, and the yields were less than 50% of theory. A recent patent has disclosed that the Perkin reaction can be carried out economically with 75% yields and fewer by-products if a catalyst is employed to accelerate the reaction. This catalyst can be the oxides or salts of metals of group 7 or 8 of the periodic system notably ferric chloride, nickel oxide, manganese chloride, cobalt oxide, etc. By using suitable ratios of reactants and controlled temperature ranges, economical yields of coumarin have been obtained. An example of the procedure given in the patent follows:

A mixture of 3 690 grams (32 moles) of salicylaldehyde, 6 190 grams (60. 6 moles) of acetic anhydride, 4 960 grams (60. 5 moles) of anhydrous sodium acetate, and 128 grams of cobaltous chloride hexahydrate was placed in a vessel fitted with a stirrer, thermometer and condenser. The mixture was heated with stirring at 150 ℃ for 2 hours during which time 440 grams of a mixture of acetic acid and acetic anhydride was distilled from the reaction mixture. The temperature was then gradually raised to 180 ℃ and maintained at 180 ~ 195 ℃ for 3 hours. The reaction mass was cooled to about 115 ℃ , diluted with about 25 liters of hot water and the mixture agitated for 15 minutes. The mixture was then allowed to stand whereupon it separated into two layers. The upper, aqueous layer was extracted with 7 liters of benzene and the benzene extract added to the nonaqueous layer. The benzene was distilled under reduced pressure to obtain 310 grams of a mixture of unreacted salicylaldehyde and salicylaldehyde acetate, and 2 920 grams coumarin having a melting point of 67. 1 ℃. The yield of coumarin was 64. 5% of the theoretical based upon the salieylaldehyde initially employed.

A recent Russian patent claims yields of 70% ~75% coumarin by forcing the equilibrium reaction toward the formation of coumarin. This is done by removing the acetic acid formed as a result of the formation of coumarin, at the same time adding more acetic anhydride to force the reaction equilibrium toward completion.

Lesson 2　Linalool

2, 6-Dimethyl-2, 7-octadien-6-ol

$C_{10}H_{18}O$

2, 6-Dimethyl-1, 7-octadien-6-ol

Mol. Weight 154. 25

Linalool is one of the most interesting acyclic terpene alcohols, both from the point of view of academic interest and of practical value. Several important essential oils such as bergamot, lavender, petitgrain and rosewood owe their valuable odor quality to linalool or its esters. For example, bergamot and avender oils are rich in linalyl acetate and may contain this ester to the extent of over 40%. The delicate and highly pleasant odor of linalool and its esters plays an important part in the composition of many floral perfumes, colognes, and various other formulations. Because of the relatively low boiling point, linalool and its esters are often employed to add a desirable lift to perfumes, and this property becomes indispensable in compositions where a light and pleasant note is desired.

It is interesting to note that linalool obtained from different sources has different odor qualities which betray its origin. This can probably be explained by the presence of small proportions of compounds which cannot be removed by the usual methods of purification.

Since linalool possesses an asymmetric carbon atom, it is capable of existing in optically active forms. It is thus a good natural source of a tertiary alcohol for the study of problems related to optical isomerism. Some of its reactions, such as its cyclization to an optically active terpineol, are not well understood and offer interesting fields of research to the organic chemist.

In view of its wide occurrence and peculiar structure, it is quite possible that there is a fundamental relationship between linalool and other terpenes, and that linalool plays an important role in the biogenesisof essential oils.

Isolation and Identification—Linalool was among the first terpene alcohols to be reported in the literature. Its empirical formula was established some thirty years later by Grosser as $C_{10}H_8O$, and its structure elucidated not long afterward as a result of thorough investigations by Barbier, Tiemann and Semmler. Bertram and Walbaum found that on chromic oxidation linalool yielded an aldehyde similar to or identical with citral. This was rather puzzling since it was already known that the primary alcohol, geraniol, gave citral on similar treatment and that linalool possessed an asymmetric carbon atom since it exhibited optical activity.

Barbier who was also devoting much time to the study of this compound, discovered that it possesses two ethylenic linkages and noted that on prolonged heating with acetic anhydride it gave the ester of an alcohol which was different from linalool. This alcohol was later shown by Bouchardat to be geraniol and the aldehyde which Barbier obtained by the oxidation of linalool to be citral.

A further advance in the study of the constitution of linalool was made when Barbier and Bouveault found that oxidation of linalool with chromic acid mixture gave not only citral but also acetone and methyl heptenone. As a result of these experiments, numerous structural formulas were put forward which were, however, later shown to be incorrect. In 1895 Tiemann and Semmler, in continuation of their studies oxidized linalool first with potassium permanganate, then with chromic acid mixture, and obtained acetone and levulinic acid. This and other evidence led them to conclude that linalool was a tertiary alcohol and that the formation of geraniol was the result of isomerization under the influence of acids.

Commercial Methods of Preparation—Commercial linalool is obtained from several sources, principally from oil of bois de rose, linaloe, shiu and coriander. Purification of linalool is not practicable because of its chemical properties, and its preparation consequently involves careful fractionation of these oils. Fortunately, linalool occurs to the extent of about 80% in the first two oils mentioned, and they therefore constitute very convenient sources of this valuable alcohol. Highly efficient distilling columns are used to effect separation of linalool from other components of the oil consisting mainly of terpene hydrocarbons and alcohols other than linalool. Since the boiling point of linalool is considerably higher than that of the hydrocarbons and somewhat lower than that of geraniol and citronellol, a good fractionating column yields comparatively pure linalool. In practice, the first distillation serves to separate the linalool from the bulk of impurities, and a second fractionation is necessary to obtain a fine grade of product. Aside from this, it is found that the various fractions of linalool possess slightly different odors, and these fractions or cuts are graded and employed for different purposes. As a rule, it is possible to identify the source of linalool merely by smelling it, since each oil contains traces of substances which are not easily removed by distillation.

Chemical methods for the purification of linalool have been proposed from time to time but they have proved to be unnecessary and impractical. It is possible to form the phthalate or the borate of linalool and then distill off the terpene impurities. Since these esters have a very high boiling point, a single fractionation is sufficient to separate the terpene hydrocarbons. Such a procedure is rather expensive, however, and is not commonly employed in view of the satisfactory results obtained by the methods described in the preceding paragraph.

Words and Expressions

linalool [lɪˈnælououl] *n.* 沉香醇, 芳樟醇

terpineol [təˈpɪniˌɔl] 萜品醇, 松油醇

bergamot [ˈbɜːgəmɒt] *n.* 香柠檬, 香柠檬油

peculiar [pɪˈkjuːliər] *adj.* 独特的, 异常的, 特有的, 特殊的

lavender [ˈlævəndər] *n.* 熏衣草属, 熏衣草花, 淡紫色

anhydride [ænˈhaɪdrɪd] *n.* 酸酐, 脱水物, 酐

petitgrain 苦橙叶, 苦橙叶精油

permanganate *n.* 高锰酸盐

rosewood [ˈrouzwʊd] *n.* 蔷薇木, 黄檀, 黄檀木, 黄檀属

coriander [ˌkɔːriˈændər] *n.* 芫荽, 胡荽, 香菜

asymmetric [ˌeɪsɪˈmetrɪk] *adj.* 非对称的, 不对称的

ethylenic linkage *n.* 烯键

isomerism [ˈaɪsəmərɪz(ə)m] *n.* 同分异构性

Reading Material: Phenyl Ethyl Alcohol

β-Phenethyl Alcohol, 2-Phenyl Ethanol, Benzyl Carbinol

$C_8H_{10}O$ Mol. Weight 122.16

Phenyl ethyl alcohol is one of the most important and most widely used of the aromatic perfumery compounds. It is interesting to note that its synthesis in 1876 by Radzisewski anteceded by at least two decades the discovery of its occurrence in oil of rose and the realization of its value as a perfumery material.

The importance of phenyl ethyl alcohol to the industry is evidenced by the amount of work which has been done and the numerous patents obtained for its manufacture on an industrial scale. Thequantity of phenyl ethyl alcohol manufactured at the present time is exceeded by only a small number of other perfumery compounds.

Because of its mild, pleasant, and persistent rose odor, phenyl ethyl alcohol finds extensive use in rose and numerous othertypes of perfume compositions. When a floral rose note is desired, it is generally employed in conjunction with geraniol, citronellol, and their esters. Its relative stability toward alkali makes it especially suitable for perfuming various types of cosmetics and soaps.

Occurrence—Phenyl ethyl alcohol has been found in many oils, the quantity being generally small except in oil of rose. In view of its occurrence in a wide variety of sources, it would appear probable that its identification in many other oils is a matter of time and further study. It has been definitely identified in the following oils.

Oil of rose obtained in Bulgaria by steam distillation contains a very limited amount of this alcohol, usually not more than 1%. If, however, the rose petals are extracted with a solvent and the concrete so obtained is steam distilled, a method generally used in France, the resulting oil may contain phenyl ethyl alcohol to the extent of 64%. This difference is, of course, due to the loss of the alcohol in the distillate waters during the process of steam distillation.

Geranium oil contains small quantities of phenyl ethyl alcohol. Geranium oil of Kivu obtained in Congo contains this alcohol in the form of esters.

Orange flower oil from the flowers of the bitter orange fruit also contains phenyl ethyl alcohol in small quantities. It has been previously identified in the distillate waters of this oil.

Oil obtained from carnations has been shown to contain about 7% phenyl ethyl alcohol.

Hyacinth flowers yield an oil in which phenyl ethyl alcohol has been identified.

Phenyl ethyl alcohol has been identified in the flowers of Michelia champacp L.

Oil of Aleppo pine needles contains phenyl ethyl alcohol as one of its constituents.

The oil extracted by means of ether from California orange juice was found to contain phenyl ethyl alcohol among other ingredients.

Chemical Properties—Phenyl ethyl alcohol exhibits many of the normal properties of a primary alcohol. It is relatively stable and on prolonged standing forms only traces of phenylacetaldehyde.

It may be dehydrogenated to phenylacetaldehyde by passing over reduced copper at 300 ℃. Phenylacetaldehyde can also be prepared from phenyl ethyl alcohol by passing it along with oxygen over silver catalyst suspended on asbestos at 300 ~ 320 ℃ and 25 mm pressure. The same result is achieved by the use of zinc oxide catalyst at 400 ~ 430 ℃ under reduced pressure.

Oxidation of phenyl ethyl alcohol with sulfuric dichromate mixture leads to a number of degradation products, including benzaldehyde, benzoic acid, phenyl acetaldehyde, phenylacetic acid, etc, Potassium permanganate oxidation, on the other hand, results mainly in benzoic acid.

Phenyl ethyl alcohol gives phenyl ethyl chloride when it is treated with hydrochloric acid at 140 ℃ or, better still, with thionyl chloride in the presence of diethylaniline.

When heated with sodium bisulfate or with a little 64% sulfuric acid, phenyl ethyl alcohol is converted to diphenyl ethyl ether.

It is dehydrated quantitatively to styrene on heating with anhydrous potassium hydroxide.

On hydrogenation with nickel at 380 ℃ the alcohol is reduced with the formation of ethyl benzene.

In common with many primary alcohols, phenyl ethyl alcohol forms a calcium chloride addition compound and this property has been used as a means of its purification. It also forms with ease many derivatives which are used to characterize alcohols.

Reading Material: Fragrance Selection—Technical Considerations

1. Chemical Changes

1.1 *Reactions with Product Components*

Most fragrance materials are alcohols, esters, aldehydes, ketones, phenols or unsaturated hydrocarbons. In the spectrum from the most stable organic substances to the most unstable ones, they occupy a middle ground. In most applications of fragrance in consumer products, chemical instability of the fragrance is not a problem. Some notable exceptions are the following:

Products containing oxidizing agents; for example, hair dyes and bleaches, laundry bleaches, heavy duty detergents, automatic dishwasher detergents, and some scouring powders. Oxidizing agents may also be formed as breakdown products in formulations containing unsaturated fatty acids, especially in the presence of metal traces and in the absence of antioxidants (rancidity in soaps is an example).

Products containing strong acids; for example, cold wave neutralizing rinses and some antiperspirants. Strong acids may also develop as breakdown products in aerosols with propellant II and water.

Products containing strong bases; for example, cold wave solutions and depilatories, hair dyes and bleaches, and heavy-duty detergents.

Products containing proteins or polypeptides; for example, protein shampoos, collagen creams, and enzyme detergents.

Products containing formaldehyde; for example, nail preparations.

The most common way of diagnosing chemical breakdown of the fragrance is by smell. Chemical breakdown usually causes changes in quality as well as in intensity. The best way to measure the degree of change, therefore, is to give panel members two samples—one fresh and one aged—and ask them to indicate the degree of odor difference between the two. A 5-point scale ranging from 0 (no difference) to 4 (extremely different) is appropriate. Asking the panel members to describe the difference may sometimes provide clues to the origin of the problem.

The chemical breakdown of perfume materials is usually a slow process. For this reason, accelerated storage tests are commonly employed, using temperatures in the range of 105 ℉ to 120 ℉. Although no firm rules can be given, a week at 10 ℉ is usually deemed to be roughly equivalent to one month at about 70 °F, and one week at 120 °F is equal to two months at 70 °F.

1.2 *Prevention and Remedy*

If fragrance stability problems are anticipated, it is a good idea to tell the fragrance suppliers which ingredients may cause problems and give them samples of the product for experimentation. When fragrance stability problems are due to one of the key ingredients of the product, the fragrance must be formulated to avoid such problems. The factors leading to chemical instability are far better understood than the physical interactions between fragrance and product. Formulation of a fragrance to avoid chemical problems can, therefore, usually be approached in systematic fashion rather than by trial and error and involves less work. Chemical problems may, however, severely restrict the range of fragrance materials at the perfumers' disposal, hence also the range of fragrances available for the product.

When the cause for the fragrance breakdown is found to lie not in a principal component of the product, but in an impurity or gradually developing breakdown product (e. g. , rancidity of unsaturated fatty acids and hydrogen chloride from propellant II) , it is best to change the product formulation rather than the fragrance the impurities or breakdown products usually have other undesirable effects in addition to that on the fragrance.

Fragrance breakdown can also be diagnosed by gas chromatography. Samples of the productare then taken at regular intervals in the accelerated aging process, extracted so as to obtain a concentrate of the fragrance that they contain and subjected to gas chromatography. If both the extraction of the product sample and the gas chromatographic analysis are conducted under rigorously standardized conditions, the series of curves obtained at different points in time provide a precise and detailed record of what is happening to the fragrance. Compared to the subjective estimation of he overall degree of difference of fragrance, which is always fuzzy, such a record is considerably more definitive and unambiguous. Moreover, gas chromatographic curves provide a means for identifying the components of the fragrance blend that undergo changes upon storage, thus permitting a diagnosis of the exact nature of the problem. Sniffing has the advantage over instrumental analysis of being less laborious and measuring more directly the critical variable— the degree of odor change the consumer is likely to observe.

2. Physiological Reactions

The record of the dermatological literature indicates a very low incidence of skin irritations caused by the fragrance in personal or household products. Still, when a product elicits a primary irritation or an allergic reaction (also called sensitization) , fragrance is often suspected first. Although frequently unjustified, this tendency is understandable. It is understandable on the part of the dermatologist because of the complexity of fragrance formulations and the fact that it is usually hard to obtain disclosure of their composition. The finished goods manufacturer may be indulging in some wishful thinking here, insofar as skin reactions caused by fragrance can be readily remedied by changing the fragrance. Reactions due to other ingredients require extensive testing for diagnosis and often substantial reformulation efforts for remedy. Last but not least, the consumer is being educated to associate perfume and allergy by the publicity about perfume-free hypoallergenic cosmetics. Though the population at large may not be widely aware of such cosmetics, the people who are susceptible to allergic reactions are.

In the past our knowledge of the effect on the skin of various perfume materials has been somewhat spotty. It has come from isolated reports spread over a wide range of journals in different languages. Some of these reports were based on inconclusive or inadequately controlled experiments or involved impure materials, with skin effects due to the impurities rather than the primary materials. Some such reports achieved an aura of authoritativeness by being widely quoted in the professional literature. Some years ago, the Research Institute for Fragrance Materials was established in the United States, whose main task is to undertake a thorough and critical compilation of all available information of the effects of fragrance materials on the skin, to complement this information by means of a testing program, and to report its findings to the fragrance industry and to government regulatory agencies. This work is still in progress. A European organization with similar aims, the International Fragrance Research Association (IFRA), has recently been founded.

The following considerations about fragrance and skin irritation should be kept in mind:

(1) Some of the hundreds of raw materials commonly used in fragrance are considered more likely to give rise to skin irritation or allergic reactions than are others. Fragrance manufacturers generally avoid using these materials in fragrancesfor products that come into prolonged contact with the skin. One of the several reasons why it is advisable to tell your fragrance supplier as much as possible about the product in which you want to use his fragrance is to alert him to any possible risks of skin irritation.

(2) Today almost all new personal products are subjected to skin or eye irritation testing. The product sample used in such tests must contain the same fragrance as will be used in the product as marketed, at the same level.

(3) A fragrance manufacturer cannot guarantee that a fragrance will never cause an allergic reaction. Allergic reactions are unpredictable; it has been said that no substance exists that never causes an allergic reaction even water has been known to evoke allergies. Moreover, the tendency of a fragrance to cause primary irritation or allergic reactions depends, not only on the fragrance itself, but also on the product in which it is incorporated. There are several reasons for this. Some allergic reactions are the result of the simultaneous action of two or more materials— one in the perfume, one or more in the base formulation. This phenomenon is called cross-sensitization. Also, certain product formulations bring the fragrance into more prolonged or more intimate contact with the skin than do others. For example, interaction between perfume or other product ingredients and the skin is far more likely in a cream or oil that is rubbed into the skin than in powders that are dusted on or in soaps that are rinsed off immediately after use. Further-

more, products that are usually applied to abraded or otherwise irritated skin (e. g. , baby products or sunburn preparations) are more likely to cause problems than those that are applied only to intact skin. A few materials are known that do not cause skin reactions in the dark but may cause them when placed on skin in direct sunlight. Here, too, we must distinguish between photo-irritants which ca use primary irritation in conjunction with ultraviolet radiation and photo sensitizers which may cause allergic responses.

Thus there is no assurance that a fragrance that has caused no problems in a number of preparations will remain problem free in a new product. The predictive skin tests before marketing should be conducted not only with the same perfume that will be used in the marketed product but also with the exact product formulation.

References

[1] 刘庆义. 化学化工基础英语[M]. 北京: 化学工业出版社, 2008.

[2] 花建丽, 陈锋, 孟凡顺. 精细化工专业英语[M]. 北京: 化学工业出版社, 2007.

[3] 刘宇红. 化学化工专业英语[M]. 北京: 中国轻工业出版社, 2002.

[4] 李维屏, 祝祖耀. 新编现代化工英语[M]. 上海: 华东理工大学出版社, 1993.

[5] 吴达俊, 庄思永. 制药工程专业英语[M]. 北京: 化学工业出版社, 2000.

[6] ZHANG S W. English for Science and Technology[M]. Xi'an: Xian Electronic Science & Technology University Press, 2008.

[7] SWALES J. Genre Analysis: English in Academic and Research Settings[M]. London: Cambridge University Press, 1990.

[8] BIRD R B, STEWART W E, LIGHTFOOT E N. Transport Phenomena[M]. 2nd ed. New York: John Wiley & Sons, Inc., 2002.

[9] QUIRK R, GREENBAUM S, LEECH G, et al. A Grammar of Contemporary English[M]. London: Longman, 1973.

[10] NEWMARK P. A Textbook of Translation[M]. New York: Prentice Hall, 1987.

[11] TYTLER A F. Essay on the Principles of Translation[M]. Amsterdam: John Benjamins B. V., 1978.

[12] CROWL D A, LOUVAR J F. Chemical Process Safety: Fundamentals with Applications[M]. Boston: Pearson Education, Inc., 2011.

[13] CHEATLE K R. Fundamentals of Test Measurement Instrumentation[M]. Durham: Instrumentation, Systems, and Automation Society, 2007.

[14] JIANG Y H. A Study on Professional Development of Teachers of English as a Foreign Language in Institutions of Higher Education in Western China[M]. Berlin: Springer, 2017.

[15] TOWLER G, SINNOTT R K. Principles, Practice and Economics of Plant and Process Design[M]. Amsterdam: Elsevier Science, 2013.

[16] CENTER for CHEMICAL PROCESS SAFETY. Guidelines for Integrating Process Safety into Engineering Projects[M]. New York: John Wiley & Sons, Inc., 2018.

[17] CENTER for CHEMICAL PROCESS SAFETY. Bow Ties in Risk Management: A Concept Book for Process Safety[M]. New York: John Wiley & Sons, Inc., 2018.

[18] MILLING A J. Surface Characterization Methods-principles, Techniques and Applications[M]. New York: Marcel Dekker. Inc. 1999.

[19] SMITH H M. High Performance Pigments[M]. New York: John Wiley & Sons, Inc., 2002.

[20] CYBULSKI A, MOULIJN J A, SHARMA M M, et al. Fine Chemicals Manufacture[M]. Amsterdam: Elsevier Science, 2001.